The Dreadful Month

The Dreadful Month

Carlton Jackson
With a Foreword by Harry M. Caudill

Bowling Green State University Popular Press
Bowling Green, Ohio 43403

Other books by Carlton Jackson:

Presidential Vetoes
Zane Grey
J.I. Rodale: Apostle of Nonconformity
The Great Lili

As co-author:

Foundations of Freedom
Challenge and Change
Two Centuries of Progress

Copyright © 1982 Bowling Green University Popular Press

Library of Congress Catalog Card No.: 82-72582

ISBN: 0-87972-205-3 Clothbound
0-87972-206-1 Paperback

This book is lovingly dedicated to my parents: My mother, Winnie Forrester Jackson; and to the memory of my father, Luther H. Jackson, who was a coal miner.

Contents

The Dreadful Month

Carlton Jackson
A Foreword by Harry M. Caudill

When the German Sixth Army surrendered at Stalingrad on February 2, 1943 newsreels astonished the world with views of almost endless lines of prisoners parading through the streets of Moscow. On and on they came, ragged, tattered, disarmed. The spectacle was overwhelming in the numbers involved and in their demoralization and misery.

In his great *Russia at War* Alexander Werth states that there were 91,000 of them—an immense legion of young men sacrificed in the folly of war. It was all very impressive and dramatic.

But war is not alone in devouring whole armies of men. Industries consume them, too, in vast and bloody myriads and no industry devours them so greedily as coal.

The World Almanac and Book of Facts for 1980 requires half a page of fine print to list the principal U.S. mine disasters that have occurred since 1855. Even so, between 1855 and 1968 the list includes only those calamities that claimed at least fifty lives. Thus, it omits the twenty-three miners who were blown to bits in Harlan County, Kentucky in 1932, the dozen who perished in the same county in September, 1943, and the thirty who were immolated on Straight Creek in nearby Bell County in December, 1945.

The scandalous truth is that in the 150 years since the railroad era began the United States have sacrificed far more men than were in those columns of straggling Germans. They perished in roof falls, were crushed by machines, and were blasted into fragments of blackened flesh by exploding methane and coal dust. At least two were permanently crippled for each of the fallen, and as many more were reduced to wheezing uselessness by industrial diseases that included pneumoconiosis, silicosis, anthrocosis, and the debilitating effects of inhaling chemical-laden fumes from burning cable insulation. These events produced more than a half million

1

widows and orphans and several times that number of welfare-supported wives and children.

Since the beginning of the coal age mine owners have been amazingly callous about the lives and health of their workmen. Graduates of Ivy League universities, members of the most exclusive clubs, heirs to immense fortunes—in short, the most fortunate of all Americans—they rigidly and persistently opposed any and all forms of compensation and safety laws. To this day the industry has yet to take the initiative in any effort to make mines safer, miners healthier, or mined lands more productive for the benefit of future generations. Of its own volition coal has chosen to be a renegade among the nation's industries.

Carlton Jackson has given us an opportunity to grasp the meaning and the shock of mine disasters that wipe out most—and sometimes nearly all—of a community's men. He has chosen to concentrate on the month of December, 1907 when hundreds of men were wiped out in a series of disasters that fell on the country with the rhythmic inevitability of drum beats. These were humble men and boys whose lives were extinguished, humble women and children who waited at the pitheads to meet the blackened corpses. They were poor, ragged, and unschooled, and few governors, congressmen or senators felt any real concern for them. As for the mine owners, the evidence indicates that they worried more about lost production than about lost lives.

It must be hoped that *The Dreadful Month* will prove to be the first of a genre that will explore in detail the coal industry's impact on the people of the mining regions, on their culture, and on the land and its creatures. Wanton deadliness is not necessary to industrial greatness and the time has come for America to civilize and curb its deadliest industry. As we shift from reliance on oil to reliance on coal, as power plants turn to coal, and as an immense synthetic fuel industry emerges to wring oil and gas from the black carbon lumps, we can learn an essential lesson from that dreadful month seventy-five years ago. If we do not the coal we burn will continue to reek of blood.

Preface

When I first read the statistics, I did not believe them. I actually fired off a letter to the Mine Safety and Health Administration seriously doubting the numbers. From 1890 to 1907, there were approximately 20,000 fatalities in the coal mines of the United States? Impossible! That's worse than any combat the armed forces engaged in during that period. Surely someone is deficient in mathematics to arrive at such a monstrous figure. Or else, the truth for some reason or other, is being greatly distorted.

Then came back the answer. No, it wasn't 20,000; it was, to be precise, 26,434. My God! And this, of course, made me want to know the overall figures. From a low of seven fatalities in 1847 to a high in 1907 of 3,242, coal mines during the period 1839 to 1977, killed 121,209 men. This was from *all* causes: fires, rock and slate falls, gas emissions, and explosions. The annual average was perilously close to 1,000 for 138 years. No amount of argument on behalf of "private property" or "free enterprise" will ever justify this kind of carnage. As this country moves into a new era in which coal comes back into its own as the chief source of heat and energy, it would be well for new safety practices to be created, because we can't really afford another "dreadful month" as in December, 1907.

By far the worst coal mining year was 1907. From all causes, over 3,000 perished. But in one month alone of that year, the last one, a climax of horror was reached as "officially" 702 miners were killed from explosions: two in Pennsylvania, one each in West Virginia, Alabama, and New Mexico territory. "Unofficially," at least a half more of that number died in the accidents, a fact that was supposedly repressed by company and government officials. The argument continues to this day about the exact number of casualties from all these explosions.

December, 1907, in coal mining history in the United States was "the dreadful month," and that is what my book is about. I have tried to chronicle the events of those thirty-one days, and to say why these terrible calamities were visited

3

upon the American people. Also, I have taken care to show that these disasters helped to change materially the social and political attitudes of the general public in this country. Not least of these changes was the creation of private welfare systems upon which the later federal programs were based, and the acceptance of yet another governmental agency, the Bureau of Mines.

December seems to be an unlucky month in any occupation where sizeable quantities of dust are present. For coal mines, the matter is explained this way: inclement weather interferes with the operation of ventilation fans placed at mine mouths to such an extent that "pockets" of gas are frequently not dispersed as quickly as they should be. They just hover around, waiting for someone to make a mistake. Also, in summer months the air blown into the mine from the warm outside is condensed to some extent by the cool interior, creating an automatic "sprinkling system," keeping the mine damp and relatively safe from explosions. In the Winter months—and for some reason, December is the worst one—the system is reversed, and no protection is offered.

In respect to December explosions, one is reminded of the grain catastrophes that attracted this country's attention in 1977. Tragic blasts in Louisiana and Texas took upwards of fifty lives, reminding the public once more that dry dust particles—whether from grain or coal—can explode. It is ironic that the grain explosions happened seventy years to the month after the disastrous events of 1907.

I have tried to present the facts in this book as truthfully as possible, and to analyse them in ways that could possibly be of value for today's generation as it goes about the task of causing companies and governments to accept their responsibilities. I hope I have succeeded—at least in part. If so, I am content.

The "perspectives" chapters are designed to give "updates" on the dreadful month of 1907, to explore the possibility of positive changes, and to report, alas, that there weren't any. Very little, indeed, was learned from the harsh experiences of December, 1907—a fact that does not speak well for the good-will and intelligence of operators and governments alike. What happened was, in a phrase, a national scandal—and it continues to this day.

For the "perspectives" chapters, I have gleaned the information primarily from press reports (which, incidentally, have always proven to be as true, or more so, than the "official" findings). Also, I have limited my discussion to accidents that resulted in fifty fatalities or more (with the exception of Scotia in 1976). God knows, I have no callous purpose in mind by eliminating all those who died in "less than fifty." Death is absolute, I am sure, to the person who experiences it, and whether it is one or 101, it has its own story to tell. Quite simply, however, if I had included *every* explosion in the "perspectives" chapters, this would indeed have become a multi-volume work. As it is, I have omitted several "fifty or more" explosions. For example, for the teens I discuss only the 1913 disaster at Dawson, New Mexico. Yet between 1908 and 1919, there were seventeen other explosions that took fifty or more lives. I hope the "perspectives" will show the necessity of vigilance in the nation's coal mines as once more they take center stage in the current energy crisis.

Many people helped me with this book, and to each I am grateful. Peter Gottlieb, assistant curator of the West Virginia University library was most cooperative. Mr. Meyer Mathis of the American National Red Cross lived up to the good name of that organization, and gave me considerable aid in collecting various papers and letters. Numerous governmental agencies put their services at my disposal, and I am appreciative toward people like R.O. Swenarton of the Bureau of Mines; Robert M. Dructor of the Pennsylvania Bureau of Archives and History; and Sam Stafford of the Mine Safety and Health Administration.

There are two people to whom I wish to express special gratitude. One is Mrs. Caroline Urban, widow of Peter, the only survivor of the Monongah blast. Mrs. Urban graciously allowed me to visit her in December, 1977, as she was preparing to celebrate her 100th birthday, happy—among other things—to have received a birthday card from President Carter. In December, 1981, she observed her 104th birthday, and at the time of this writing (March, 1982), she is anticipating her 105th. The other person is Mr. Lester E. Trader, who granted me an interview in May, 1978—just before his death in July. Both of these direct links with the past give my study, I believe, a sense of humanity that otherwise would be lacking. I feel extremely fortunate to have had the opportunity of coming into contact with these two fine people.

Numerous individuals at my University, Western

Kentucky, helped in the preparation of this book. My friends and colleagues, Professors Helen Crocker and Richard Troutman read all or portions of the manuscript and gave valuable advice. Research grants from WKU promoted the work considerably. To one and all: thank you.

    I thank also my family: Pat; Beverly, Steven and Colleen; Hilary; Dan, Grace, Travis and Megan; and Matthew.

Western Kentucky University   **Carlton Jackson**

Chapter One

The Dreadful Debut

After a day of intermittent snow showers, a beautiful twilight descended on the little towns and villages that dotted the valley drained by the Monongahela River. Benevolent but fading sunlight sent rays through the chilled air, casting dappled images here and there upon miners' cottages, or dormant trees. The rays' sweep took in the deep, yawning black holes in the ground, vivid testimonials to the rape of this valley by the swiftly growing, all encompassing coal industry.

The citizens of these hamlets—which included West Newton, Belle Vernon, Greensburg, Charleroi, Fairhope, and Fayette City—stirred themselves away from the beauty of this day, and prepared for the evening's activities. Many dressed for church, for this first day of December, 1907, was also the Sabbath. Others called on friends and relatives, as Sunday was really the only day of the week they could socialize. Still others simply sat at home, enjoying nature's handiwork, quietly reflecting upon the world in which they found themselves.

A portion of Fayette City, however, was not privileged to view the sunset. They were a part of the two hundred men who worked for the United Coal Company's Naomi Mine. They went into the pit at 4 that afternoon, expecting to work a full shift—until midnight....

One of their number was Joseph Robish, a waterman whose job, among other things, was to keep the floors dampened to prevent undue accumulations of coal dust. He was known affectionately to all his friends as "Pumper Joe." It was a common practice for him to leave half way through a shift, once he completed his rounds. Today was no exception, particularly since so small a force had shown up for work that afternoon.

One of "Pumper Joe's" loves was his pipe. He, along with some of his buddies, frequently lit up, regardless of where they

7

might be, disregarding the pleas of their brethren that they refrain from the practice. They always laughed away such requests, pointing out that Naomi was not gaseous, and there was nothing to worry about, a statement that had been flatly denied for the past year by the Pennsylvania Department of Mines.

On his way out, "Pumper Joe" took a short cut through some inactive areas of the mine. These were not worked out, abandoned places, but sites from which at the moment no coal was being extracted. According to later evidence, and professional surmising, "Pumper Joe" paused in one of these places to take a smoke. After filling his pipe with tobacco, he took off his cap and, to get a light for his pipe, removed the lid of his sealed lamp.... At 7:26 that Sunday evening, so later testified the engineer of the power house, a circuit breaker blew, followed almost immediately by a huge explosion.

Fayette City's residents poured out of their homes and churches and, though night had now fallen, looked automatically toward the Naomi Mine. They instinctively knew what had happened. All at once, it seemed, they rushed toward the scene, some taking short-cuts across the railroad trestle that spanned a ravine between the village and the mine. They accumulated in such large numbers around the mine's mouth that rescue workers could not get into the area. A large proportion of the on-lookers were foreigners; consequently, they could not understand the officials' orders that they stand back and disperse. Instead, they screamed and shouted and wailed and moaned, and a few helplessly beat their breasts and pulled at their hair. Some prayed loudly that their relatives inside the mine be found safe; while others, perhaps remembering that swearing is sometimes considered to be the highest form of prayer, cursed loudly.

After three hours of bedlam, in which no order whatever was evident, mine officials and local government authorities were forced to call for help from several surrounding towns. It was possible, therefore, only by midnight to rope off an area to keep survivors and spectators at a distance, so the work of rescue could begin. State police reinforcements came in immediately from Mt. Odin and Charleroi. Several hours later, they were augmented by forces from Monongahela and Washington, Pennsylvania. By daylight, on December 2, special police numbered about thirty.

Throughout much of the state and nation, this build-up of police power was later considered as a vast over-reaction by the company and government officials. The latter, however, asserted that the crowds had to be controlled so the rescuers could get in. They also pointed out that "agitated foreigners" were several times just on the verge of a riot—the authorities feared that the more emotionally stricken of them might try to charge into the mine itself, to recover their loved ones. A large security force was, therefore, vital. Despite these disclaimers of "bias against foreigners," and irresponsibility, United's actions at Naomi actually helped to provoke a feeling of reform throughout America, a feeling that, helped along by other disastrous events, was indeed widespread by the end of December, 1907.

Rescue operations were led by a state mining inspector, Henry Louttit, and Naomi superintendent of two months J.D. O'Neill. The rescuers, who came from near-by mines, were seriously hampered by continuing rock and slate falls. Many times an area cleared for passage before being properly bratticed, would again cave in. This caused forward movement to be very slow, making it ever more likely that no survivors would be found.

The greatest danger to the rescue workers, however, came from the blackdamp or, as some called it, afterdamp. This is a mixture of gases, consisting primarily of carbon-dioxide and nitrogen, which acts directly and quickly on the red corpuscles of the blood, very frequently with fatal results. It is produced by a large explosion or fire that occurs in a relatively contained area. The blackdamp is irrespirable, causing those who come into contact with it to choke and suffocate. One of its side effects is pneumonia. The searchers working their way through Naomi's passageways—three shifts a day, thirty men to each shift—had to be relieved every half hour because of congested lungs and exhaustion brought on by the blackdamp.

This fact alone made it obvious to everyone that none of the entombed miners would be found alive. Thurston Wright, assistant secretary for the United Company, made an official statement to that effect, hoping this would cause the waiting crowds to leave the premises. Wright's remark, however, had the opposite effect. It made those already on the grounds want even more to go in after their loved ones, and it was a virtual invitation to every sight-seer in the area to come and gratify his morbid curiosity. (And, indeed, the "traction" companies of the Monongahela Valley did a landmark business during the next few days, bringing many people to the site, producing near-riot conditions so often that the authorities thought seriously of calling in the National Guard.)

As the rescuers slowly and painfully worked their way into the depth of Naomi, they discovered that the explosion itself accounted actually for very few deaths. This was explained by most of the bodies not being maimed or mutilated. On the contrary, the miners must have heard the blast, and realized at once that they should make an exit. Many had fallen as they walked along a haul-way. Others were found in sitting positions, some with dinner buckets in their hands; others in a stance of prayer. Undoubtedly, they realized as they headed for the outlets that they were doomed by the instantly spreading blackdamp. One person did manage to get to the surface, but he died within a few minutes because his body was full of the deadly poison.

In the midst of the rescue attempts, a spirited controversy erupted between Louttit and O'Neill over methods of operation. One of the earliest efforts at rescue was to force an entry to the mine's main passageway. A long, spiraling staircase ran down from the surface for several feet into a cave-like area. At this site, the main passageway started. It had been blocked by tons of debris, and by wrecked cars. This had caused the early rescue work to be conducted at other passageways. Louttit, however, speculated that most of the miners would try to escape via the main passageway, so he set men to work there, trying to make a clearance. O'Neill, stating that the constant rock and slate falls would probably produce more casualties, stopped the work. This angered Louttit, and he countermanded O'Neill's order. When the way was finally cleared, fifteen of the thirty-four victims were found right at the blocked entrance, while most of the others were apparently making their way to it. O'Neill's actions caused a further mistrust of United; the general consensus was that if he had not stopped the clearing efforts at the main entry, most of the miners, if not all, would have come out alive. This was conjectural; yet, the impression remained for many months. This was not the only quarrel between Louttit and O'Neill that, as we shall see, had serious repercussions.

As the bodies were removed, they were brought, not to the

surface, but to the area of the main passageway, still some three hundred feet down. Superintendent O'Neill was largely responsible for this action. He deliberately gave incorrect information on the length of time necessary for the recovery operations, hoping by this to cause the large crowds to leave. When this ploy did not work, he threatened those present that he would not allow any bodies to be surfaced while there were still large gatherings on the mine grounds. This cruel officiousness did not in the least help United's image in the minds of the American people.

A more reasoned explanation for not bringing the bodies immediately to the surface was offered by the coroner of Fayette County, Dr. A.S. Hagan. He said he wanted to make them presentable before allowing their families to view them. It was, therefore, three days after the explosion before the victims were finally brought to the surface. Even then, Dr. Hagan insisted that they all be placed in the pumphouse, which served as a temporary morgue, until he could make final dispositions. In the meantime, the wives and mothers and children and parents had to keep waiting.

And, in their waiting, rumors and stories and speculations became rife. For one thing, it was widely reported that between forty and fifty men, not thirty as claimed by the company, had gone into Naomi that afternoon. This statistical controversy between the miners and the officials seemed to characterize every mine disaster in the country: the miners always arguing that the companies suppressed the real figures.

Much conversation swept through the crowds about how a few months ago, an explosion had occurred at Naomi, badly burning a "trapper" boy. (A trapper boy was generally eleven or twelve years old. As the *UMW Journal*, June-July, 1976, explained: "Doors underground were used to route air for ventilations, and the trapper boys would open and shut the doors whenever miners and mules needed to pass through with a load of coal.") The company at that time had been urged to correct various problems, but to no avail. Also, there had been a new air shaft under construction here for a long time, to relieve some of the pressure on the Robinson fan which had never been strong enough properly to ventilate all the passages in the mine. Unfortunately, the shaft had received a very low priority, and work on it had always gone at a snail's pace. The air shaft, so said one widespread opinion, would have prevented the explosion, because it would have taken up the weather-caused slack in the fans that allowed unusual amounts of gas to accumulate. This general belief, too, was speculative, but it was one more impression that lingered on for many months.

The story of two black men, Major Walker and Henry Hill, got a lot of currency from the crowds. Walker and Hill were partners at the Naomi Mine and Walker had wanted to work the Sunday shift to get some extra money, and had asked Hill to join him. "No-sir," was Hill's reply. "I'm a church member, and a Baptist! I ain't gonna work for no man on Sunday." The refusal angered Walker, and he stalked away from his friend's house. On Monday, however, Walker went back to Hill's house, and tearfully thanked him for saving his life. Hill responded, grinning widely, "Didn't I tell you it was against good principles to work on Sunday?"

Even as the crowds waited, burial arrangements were looked into. United President, W. S. Kuhn, announced that he would allot forty dollars to each victim for burial expenses, as several ministers arrived to conduct the necessary rituals. Father A. J. Wiggle of St. Leonard's Church in near-by Monessen, was the first clergyman on the scene. He was soon joined by the local ministers and by the Reverends A. Hasanyi of Homestead and A. Kalarski of Pittsburgh. These latter two checked off the names of their countrymen (Hungarian and Polish) in the mine, and reported that the official figures were incorrect—that at least fifteen more than counted were still in the depths.

By December 5, most of the burials had taken place, and the physical signs of the great disaster had been erased. The emotional stresses continued, however, and now a new controversy rose up between Louttit and O'Neill. The latter claimed that the State Department of Mines had never given any notices about the dangerous conditions in Naomi Mine. Was he correct?

Here is a letter, November 9, 1906, from Francis Feehan, President of District Five, UMWA, to Louttit: "I have a letter in my possession coming from miners at Naomi Mine near Fayette City which is in your district saying that the mine was unsafe, and there is constant danger of a serious explosion unless much needed improvements are made in ventilation. I think it is my duty to notify you to this effect as the miners fear they would be discriminated against if it became known that they had written to you concerning this matter. They also complain that they are not properly supplied with timber to keep their places in safe working condition. By consulting some of the English speaking miners who are working at that mine, they can perhaps describe the dangers in a satisfactory way."

Here is Louttit's response, November 10, 1906, to United official, John Underwood: "Please have your firebosses make a complete examination of your Naomi Mine on Monday morning, November 12. If there is any firedamp present that can be detected anywhere by the ordinary safety lamp, let the place or places be fenced off, and the law strictly complied with, without any reservations whatever. Report the matter to this office. Ventilate mine according to law. See that the miners are supplied with timbers."

Here is a letter from Louttit to E.J. Morris, another United official, November 13, 1906: "Complaints reached this office on the tenth [from the UMWA] that the condition of your Naomi Mine was such that there was a possibility of an explosion of gas in the same." In this letter, Louttit stated that he had personally inspected Naomi on the 12th, and found that gas was copiously being formed. It could be detected by an ordinary safety lamp. This, in the context of Naomi's inadequate ventilation system, created a condition fraught with danger.

To each of these letters, the various superintendents and managers at Naomi gave assurances of compliance—and they continued to do so throughout the year 1907. It is clear, therefore, that Superintendent O'Neill's charge of negligence against Louttit was hasty and misdirected.

After the explosion, Mr. Feehan quoted his letter of November, 1906, and remarked: "There can be no doubt that the disaster at the Naomi Mine on Sunday evening was primarily due to negligence." (Just who was negligent—an individual miner or the coal company—he did not say.) "There can be no doubt as far as our organization [the UMWA] is concerned, we still demand and insist on a most rigid and searching investigation being made into the causes which led to the snuffing out of the lives of 34 miners last Sunday evening."

And—as late as the middle of January, 1908, it was still necessary for Louttit to write to the *new* superintendent at

United, James G. Henderson, that Naomi was *still* gaseous, *still* deficient in ventilation, *still* had unusually large deposits of coal dust, and *still* there was the very real danger of another explosion. And, so it went, with little or no impact upon the big companies until the slaughter reached a point that public opinion would no longer tolerate.

Coroner Hagan's jury rendered its report in mid-December, as did the State Mine Inspectors. The only thing for certain that came from the two findings was that an explosion occurred on December 1, and killed thirty-four men. Whether Joe Robish must continue to bear the onus of allowing his pleasure of pipe smoking to snuff out his own and thirty-three other lives, or whether an electric spark detonated the gases, neither the Coroner's Jury nor the Inspectors could conclude.

Officially, the jury's report said: "We find Joe Hagardis, Frank Riske, *et al*, came to their deaths as a result of an explosion of gas and dust in the Naomi Mines...This gas seems to have accumulated in insufficient ventilation, and we believe ignited from the arching of the electric wires or an open light at some point not definitely located."

The inspectors agreed in general with this judgment, but added that the superintendents and mine foremen at Naomi when the explosion occurred were inexperienced and indulgent, having been there for only two months. Also, the inspectors claimed, if the fireboss had done his duty that Sunday afternoon, no miner would have entered the Naomi pits. It was the fireboss's job to detect and deal with gases in any appreciable amounts. Apparently, no check had been made for that day, for no record book of it was found.

Louttit, along with five other state bituminous inspectors, recommended that the Naomi officials install and maintain a sufficient water system to keep the dust wetted down. They also urged that any use of explosive powders be carried out by men especially employed for the purpose, and they completely condemned open lights in *any mine*, whether or not it was reported to be gaseous. Above everything else, however, stressed both the coroner's and the inspectors' reports, the individual activities of miners and foremen alike must constantly be supervised, because it became more and more clear that the Naomi disaster happened because of negligence on some miner's part (if not "Pumper Joe," then someone else), and indulgence by some foreman. As the Pittsburgh *Dispatch,*

December 15, editorialized: "Obviously, if we are to have safety in the mines, we must have a general revision of the practice, and extraordinary care, of selection of men who manage the workings. Scientific theories about gases and explosive mixtures are all right, but any miner of experience will tell you there's a great difference between superintendents. The human factor looms large."

And indeed, the human factor did loom large in more ways than one. Throughout the vicinity, citizens opened their hearts and their pocketbooks to the Naomi sufferers. Wagon loads of food and other provisions were sent to alleviate hunger, and to clothe survivors and keep them warm.

While the Naomi blast may have brought out the generous attitudes of the citizenry, it also revived a long-smouldering debate about unionism in general and the United Mine Workers specifically. Labor unions had never been popular in the United States, a country dedicated to the sanctity of private property.

Unionism among coal miners can be traced back at least to the 1840s, a decade that produced a laboring consciousness in the United States. The miners—primarily those in the anthracite fields of Pennsylvania joined with carpenters and textile workers in trying to get higher wages for their services, and shorter work hours. Beyond their laboring brethren, however, the miners sought safety regulations in their places of work as much as other improvements.

The miners and their co-unionists were usually the objects of wrath from the American public and, most assuredly, from the courts. Americans venerated the concept of private property to such an extent that any person or movement who even remotely threatened it was suspect. Labor unions, in their agitation against privately owned businesses, were placed into the category of "socialists" out to destroy the free enterprise system. The courts usually branded union activity as conspiracies in restraint of trade.

The American Miners Association, with its theme song of "Step by step the longest march can be won, can be won; drops of water turn a mill, singly none, singly none;" was formed in Illinois in 1860 to represent the bituminous interests of the Midwest. The Association, however, had little more success than the scraggly efforts of the 40s and 50s. It still had to fight a public opinion that grew increasingly hostile with every union-operator confrontation. Particularly did the public and government view with alarm a strike, or even a threat of one. This was considered an affront to the dignity and well-being of the country as a whole.

Labor organizations, like so many other institutions, were changed by the Civil War. Thousands of immigrants—primarily from England, Wales, Scotland, and Ireland—were hired by mine owners to replace the American workers who had gone off to war. The newcomers had been coal miners in the old country, and they brought with them their organizational skills. The sixties in some respects at least were more successful for American coal miners than previous decades, as the British influenced American Miners Association won various battles with the owners, primarily on local levels.

The panic of 1873 destroyed the American Miners Association. The entire decade marked a steady decline in the fortunes of the American coal miner. For one thing, wages were tied to prices, and since over-production was endemic to the industry, miners found that they earned less as the price of coal fell. This was especially true when they were required by operators to dig tons reckoned at 2,400 pounds. Their living expenses—particularly housing and food, which they were frequently required to obtain from the company itself—continued to rise.

In the 1880s, miners' groups were often attached to the Knights of Labor, whose object was to represent every working man in America. Adverse publicity caused mostly by strike violence weakened the Knights, and their client organizations. They were further decimated by competition from the newly formed American Federation of Labor, to which was connected a mining association called the National Progressive Union of Miners and Mine Laborers.

The Knights' miners and the Progressive Union quarreled with each other, being affected as were rival unions in other fields, by the question of political affiliation. Should they pursue their interests through an already established political party, or should they help form one of their own? These infights in the general field of labor delayed unionism of workers, sometimes by several years.

The miners ultimately realized that increased wages, shorter working hours, safer working conditions, and various social programs would be impossible while they fought among themselves. Therefore, in the late 80s, negotiations began between the competitive groups, aimed at some form of consolidation.

Thus it happened that in Columbus, Ohio, on January 25, 1890, the United Mine Workers of America was formed. Its preamble read in part:

> There is no fact more generally known, or more widely believed, than that without coal there would not have been any such grand achievements, privileges and blessings as those which characterize the nineteenth century civilization, and believing as we do that those whose lot it is to daily toil in the recesses of the earth, mining and putting out this coal which makes these blessings possible, are entitled to a fair and equitable share of the same. Therefore, we have formed the United Mine Workers of America...

The preamble went on to state the goals of the new union: earnings compatible with dangers; weekly payments in lawful money, rather than company scrip; safety regulations; an eight hour work day; an end to child labor; proper weighing of coal dug by a miner; legislation abolishing use of Pinkertons or other forces in labor strife; and collective bargaining.

Tall orders, these; and the quest for their fulfillment made labor history over the next several years. As expected, the owners and operators, the court systems, and much of public opinion, squared away against the coal miners.

It took many years, for example, even to make a dent in the practice of child labor. Lads of nine and ten were used as "trappers" inside the mines, and as "breaker boys" (who picked out rocks and other impurities as the coal came by them in long chutes) on the surface. Many of these youngsters were admitted to membership in the UMW, causing one operator to complain, "the unions are corrupting the children of America by letting them join their illegal organizations." A miners' representative answered, "If the children had not been at work in the mines, they could not have joined the union."

Child labor in America began to dissipate as each state passed compulsory school attendance laws. Then, in 1916, Congress enacted the Keating-Owen Child Labor Act, which forbade, among other things, interstate shipment of mine and

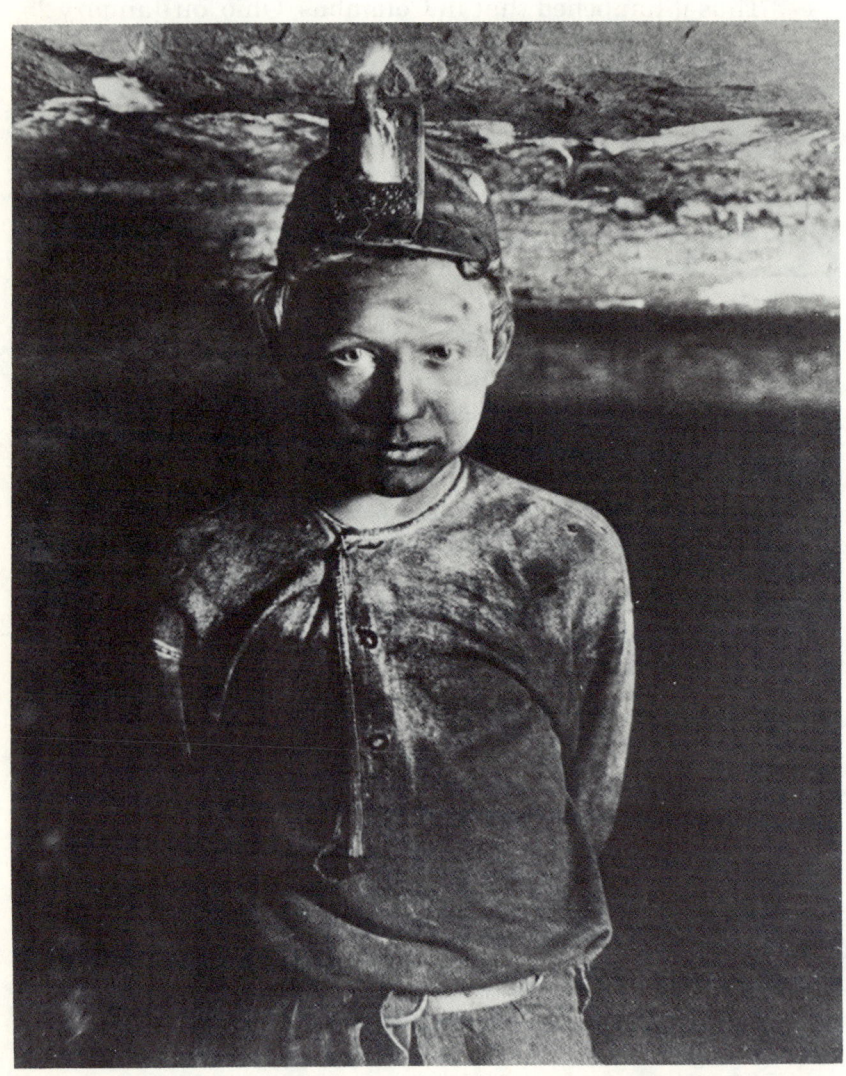

A young coal miner, ca. 1910. Photo courtesy, UMWA.

quarry products which involved the activities of children under sixteen. Two years later, the Supreme Court invalidated this Act, but Congress retaliated by imposing a ten percent tax on all interstate commerce produced in whole or in part by children under fourteen. Then in 1938, in the all-embracing Fair Labor Standards Act, employment for those under sixteen again came under federal regulation.

Gaining the eight hour work day was another up-hill climb for labor unions. In 1898, after a coal strike that paralyzed much of the operations in Pennsylvania, Ohio, Indiana, Illinois, Kentucky, and Tennessee, the "Interstate Agreement" was reached between the operators and the UMW which, among other things, established the eight hour work day. This was such an astounding victory for the coal union that membership rose immediately by one thousand percent.

Also, in 1889, one of the two men generally hailed as "super-heroes" of the coal mining movement, became President of the UMW. His name was John Mitchell (the other "hero" was John L. Lewis). Mitchell was twenty-eight when he assumed the presidency. His greatest test, perhaps, was in 1902, when the anthracite strike occurred in Pennsylvania. An operator from that state, George F. Baer, issued the notorious statement that confirmed the "Social Darwinian" stance of owners in general:

> "The rights and interests of the laboring men will be protected and cared for—not by labor agitators, but by the Christian men to whom God in His infinite wisdom has given control of the property interests of this country."

This arrogance helped to turn public opinion in favor of the strikers— at least, on that one particular occasion. President Theodore Roosevelt finally forced a settlement between the workers and the operators. This did not reflect, however, a recognition of the workers' union by the owners; rather, it demonstrated the forcefulness of the presidential office.

Another of labor's problems was the unlimited immigration policy adopted by the federal government. During the 1890s, newcomers from abroad, principally central and southern Europe, increased dramatically. In general, labor organizations scorned the emigres, for they were frequently used to undercut wages and to break strikes. Their lot was no better, either, with management. Baer, whose insolence was matched only by his garrulity, said this about them: "They don't suffer. Why, they can't even speak English.!"

UMW President Mitchell, however, was far-sighted enough to gain the foreigners' good-will, and to assimilate them into his union. His phrase, "The coal you dig isn't Slavish or Polish or Irish coal—it's just coal," had an important impact on his membership. Consequently, there was less friction between the native coal miners and foreigners than in several other industries. This fact, however, was not so easily recognized at various times during the dreadful month of December, 1907.

Thus, when Naomi blew up on December 1, its greatest tragedy was, of course, the loss of thirty-four lives. Beyond the casualties, however, the explosion revived an interest and controversy in the advantages and disadvantages of unionism. The intensity of the debate, however, was short-lived, for events in Monongah, West Virginia, dictated that a purely humanistic response be rendered.

Sources

Annual Report of the Pennsylvania Department of Mines, 1907

Arble, Meade. *The Long Tunnel: A Coal Miner's Journal.* New York: Atheneum, 1976.

Cohen, Sandford. *Labor in the United States* (2nd ed.). Columbus, Ohio: Charles E. Merrill Books, Inc. 1966.

Greensburg (Pa.) *Daily Tribune,* December 4, 5, 1907.

Greensburg (Pa.) *Press.* December 4, 11, 1907.

McCarthy, Justin. *A Brief History of the United Mine Workers of America.* Washington, D.C.: UMW, undated.

Morris, Homer Lawrence. *The Plight of the Bituminous Coal Miner.* Philadelphia: University of Pennsylvania Press, 1934.

Peterson, Bill. *Coaltown Revisited. An Appalachian Notebook.* Chicago: Henry Regnery, 1972.

Pittsburgh *Gazette-Times.* December 2, 3, 4, 5, 1907.

Rayback, Joseph G. *A History of American Labor.* New York: The Macmillan Co., 1959.

Roberts, Peter. *Anthracite Coal Communities.* New York: Arno Press and the New York Times, 1970.

United Mine Worker's Journal. January 16-31, 1974; June 15-July 15, 1976.

Chapter Two

Black Friday in Monongah

The day in December broke crisp and clean on that Friday in Monongah, and in a short time the chilled, silent air was rent by hundreds of hurrying feet. The coal miners were glad to get back to work. After all, it was already the sixth of the month—well into the Christmas season—and they could do with a bit of extra money right now. Times had not been exactly good during 1907; in fact, newspapers had frequently headlined stories about "depression," and told how certain companies had suffered from the economic privations. The miners knew this all too well. Why, just the day before, their mines—Numbers Six and Eight—had been closed, and this was one of many such days during the recent past. So it was good—really good—to get back to work.

As one of the miners walked along the railroad track toward Number Six, a large black cat came up and began to nuzzle around him—running through his legs, hindering his progress. Laughingly, he looked at the cat and saw that it was one-eyed. Strange, he thought, there doesn't seem to *be* a place for a second eye....

As each worker neared the pit of his mine, he stopped and had his name called off from a prepared roll. Then he collected his open-air carbolic lamp, and a supply of black powder and fuses. Thus equipped, he put his lunch pail into his right hand, swung his pick with shovel attached through the handle onto his left shoulder, and began his descent into the depths.

These were good places to work—Number Six, opened in 1899; Number Eight, in 1905. The ventilation, for one thing, was far in advance of any other mines in the area. A Capell fan positioned at the mouth of the pit did the job for Number Six, while a monstrous Lepley blew into Number Eight. The combined work of the two fans kept both mines—connected to each other by a passageway some three thousand feet from the entrances—at about sixty degrees. The only problem was that

21

sometimes the rush of air caused by the fans got up to forty miles an hour, occasionally blowing out the miners' lamps. This was only a small nuisance, however, compared to the comfort that these giant machines produced. Both had run all day yesterday, though the mines were closed. This had kept the comfort level high, and the mines dry.

Once inside the mines, the men went to their designated "rooms," and began extracting the precious ore from the earth. Most of the coal right now was to be found about one and a half miles back from the entrances, and that is where most of the miners were. Picks and shovels flashed deftly through the crevices, while sounds of shots being fired set up an almost continuous bang.

What was a shot? The miners, to get the coal out faster, would bore a hole of varying depths in the face of the seam. Then they would pack black powder at the bottom of the hole, and run a fuse from it. They preferred to tamp the fuse with clay because that would confine the explosion to the drilled hole, and cause the coal to crumble, making it easy for the picks and shovels to do their jobs. But if they did not have clay, and were in a hurry, the miners used anything available—dirt, mud, coal-dust, etc. If the shot were not tamped properly there was always a possibility of a sheet of fire shooting from the hole upon ignition. It was not unusual for fifteen or twenty shots to go off at the same time throughout the mines. The men on this morning were in a hurry; therefore, the frequency of shots was greater than usual. And, according to one man, the accumulated dust in the mines was, in places, hip deep.

A few hours after the work started, three men—two Italians and one American—found good reasons to leave their work places. One had forgotten his tobacco pouch and a shift without a chew was unthinkable. Another's lamp had unexplainedly gone out and he wanted the smithy to check it. The American had gone into the pits that morning still reeling from a hard night's drinking. He found an unexpected amount of money in his pocket, and his reaction was: "why work when you can drink?"

The three found the bright, welcome, sunshine gleaming down upon the little industrial hamlet of Monongah, the largest "dot" in the thirty mile span of mines from Fairmont to Clarksburg. The settlement, known first as Briartown, was a typical example of early twentieth century industrial amalgamation. Its earliest developer was Johnson N. Camden, who changed the name and created the Monongah Coal and Coke Company which, in time, became one of the largest operations in the state. Then the Watsons of Fairmont, C.W. as President and S. L. as Treasurer, formed the Fairmont Coal Company, absorbing Camden's Monongah properties. The Fairmont Company then became a part of the Consolidated Coal Corporation, which the Watsons bought from the Baltimore and Ohio Railroad in 1903.

The steel coal tipple at Monongah, said to be the largest in the world, was the most common landmark for the 6,000 residents of the village. The mines were located on the west fork of the Monongahela River, about six miles south of Fairmont. The Baltimore and Ohio Railroad ran along the eastern bank of the river, while the Clarksburg Traction Line occupied the west. Numbers Six and Eight mines were on the eastern side of the west fork, stamped like two great black eyes into the mountain that rose up from the river. Interspersed into this mountain, and almost directly on top of the mines, were the dwelling places of the miners and their families—crude, two-story frame-houses—where overcrowding was endemic. The mines were connected to the west bank by a steel pier bridge which, of late, had been in need of repair. Workmen had taken to using dynamite to dislodge some of the old piers; thus, a few detonations every day were not alarming to the citizens of Monongah.

Combined, the two mines covered about seven hundred acres, of which only sixty-two had been completely exhausted of the "black gold." One hundred seventy acres were currently being mined, leaving an actual four hundred sixty-eight yet to be developed. Little wonder, then, that Monongah was a workers' magnet—there was enough employment here for at least the next twenty years.

Will Jenkins, the blacksmith, returned the lamp to the Italian miner, smiling a bit at the simplicity of its repair, and turned back to his major task of the morning of shoeing a mule. He saw the ears of the beast suddenly go back and fit almost flat against its neck. Simultaneously, a loaded trip of cars flew by his shop, *but* in the wrong direction! It had apparently been pulled to the "knuckle," the ridged turning point between the pit and the outside, and then a coupling pin had broken, plunging the entire cargo back into Number Six. Through

Smithy Jenkins' mind there raced this thought: "My God, there are eighteen cars on that trip, and each of them has two tons of coal." He knew that to the bottom of Number Six slope, it was seven hundred forty feet, on a nine percent grade. That trip would be traveling mighty fast when it hit bottom. Jenkins ducked. . . .

Work had gone well that morning in Number Eight. Floating out of the depths were Italian arias, Polish polkas, and American folksongs, as the happy miners hummed, whistled, and sang their way through the seams. Intermingled with these joyful notes, though, was the never-ceasing cacophony of fired-off shots. As time passed, the men became increasingly careless.

For example, in the main west entry a hole was drilled seven feet into the coal. When the shot went off, coal was thrown out thirty feet into the underground chamber, whereas a proper shot would have contained it in and around the seam itself. Moreover, only five feet of the solid coal were dislodged; yet the shot was for seven. This produced an uncommon amount of fire as the shot exploded, fire that caused the fine particles of dry, floating coal-dust to sizzle and crackle.

The dust was caused by endless human boots and mule hooves breaking up the small pieces of coal on the floor. Some miners were more careful than others in handling this dust. Often, they swept the area clean before firing a shot. Too frequently, however, they ignored this simple safety precaution, particularly if they were in a hurry as they were today. Usually a watering crew came through both mines several times a day and sprinkled the dust. This procedure had not been carried out the day before, since the mines were closed. The fans had continued running, though, making the dust drier and drier.

While this shot was being fired in the main west entry, similar action occurred in room fifty-one of Number Eight. This time, fully four feet of the hole remained, and the coal from the shot flew directly across the room, some twenty feet away. Also, at a place the miners called "third left entry off the second north face," inexperienced and amateurish hands were having a go with the black powder.

Each time a shot was fired, the men ran for protection, more from the flying coal than from the explosion itself. That is what the workers in the crosscut between rooms twenty-one and twenty-two did. They intended to use a quantity of black powder sufficient to dislodge enough coal to shovel into the cars for the rest of the day—and it was only a few minutes past ten. What they did not realize was that the amount of powder they proposed to use was enough to volatilize fifty-two tons of coal—to produce 225,000 cubic feet of explosive gases....

. At 10:25 that morning, Christiana Ceredili, an Italian woman who lived on Hill Number Three, on the opposite side of the river between Mines Six and Eight, busily swept off her front porch. Suddenly there was an explosion. "Workmen on those bridge piers," was her first reaction. Then she looked up and saw smoke coming from number Six; seconds later, smoke poured out of Number Eight.

. H. L. Sloan, tippleman at Mine Number Two, also across the river, saw it too. Out of Six came "bright yellow smoke," followed immediately by a concussion in Eight.

. Luther Toothman, carpenter for Six, temporarily on loan to Two, saw the smoke from Six. It was "dirty white," he said.

Entrance to Mine #8; Monongah, W. Va. after explosion, December 6, 1907

. George Petticourt, outside foreman at Eight, felt a rumbling and shaking of the ground. He saw beams and heavy timber in the air, being blown out of the mine's opening. A few seconds later, he remembered, an explosion occurred at Six.

. Joe Newton, a black man, was standing fifty feet from the entrance of Eight. He was blown into the air, suffering the loss of his right eye and two of his fingers. He was one of only two men directly connected with the explosion who escaped with his life.

. Jimmie Rogers, a miner from Scotland, felt the trembling, sensed the rumbling, and heard the explosion, and knew at once why he had not reported for work that morning. He had not been able to pinpoint his reason for staying home until now. A few years ago he had worked at West Virginia's Newburg Mine. He awakened one morning and a "voice" told him to stay home. An explosion had occurred that day at Newburg, injuring and killing dozens of men. That same "voice" had spoken to Rogers this morning.

. George Byce was pulling some cars up track toward Fairmont. He recalled that he had smelled a bit of gas along the headings earlier in the day, especially in Number Eight.

. William H. Byce, engineer, stood in the fan house near Eight. He was injured by flying debris. He died later in a Fairmont hospital.

. Otto Smith, tippleman at Eight saw, as George Petticourt did, Number Eight disgorging its contents, and remembered immediately the complaints of some of his fellow workers that the mines were not safe.

. John Talcott, a shipping clerk, was talking on the telephone with a friend at Number Six tipple. Suddenly, for no immediately explainable reason, they were disconnected.

In only a short time the residents of Monongah, and even of Fairmont six miles away, knew that something dreadful had happened. Sidewalks in Monongah buckled and broke, streets opened in fissures, buildings shook and some of the smaller and weaker ones actually collapsed. All over the entire general area, including Fairmont, a pall of soot hovered in the air, finally falling, creating a thick coat of sludge over the river. Though the citizens were horrified, they were not surprised. The history of deep mining for bituminous coal in the United States had taught some bitter lessons.

A reporter for the Pittsburgh *Dispatch* described the scene:

What had first seemed like distant thunder, in a few
seconds was transformed into a roar of a thousand
Niagaras. Like an eruption of a volcano the blazing
gas rushed to the surface, and vomited tongues of red
flame and clouds of dust through the slopes. The
thirty foot fan [at Number Eight] which supplied the
fresh air was lifted like a toy and wafted across the
river. Poor little Charles Honaker, fifteen years old, a
trapper, with clothing ablaze, literally a human
torch, was enveloped in the fiery torrent. Several men
who were in the mine near the entrance were likewise
carried in the claws of death and strewn in the pit
mouth. Monongah mines have blown up a thousand
men.... The words were repeated from mouth to
mouth—but everyone knew it was useless to burden
human tongues with the message of the tragedy.

The extent of the accident was forcibly brought home when
a trainload of Fairmont physicians disembarked at
Monongah. Among them were the pride of the West Virginia
coal-fields, all of them M.D.'s: Hal Hall, E.W. Howard, E.P.
Fitch, J.J. Current, and Hugh Carr. Sorrowfully, however, they
had no job here to perform.

A coal mine official who arrived with the physicians
stumbled blindly around the area moaning over and over: "My
Lord! My Lord! There are over six hundred men in those
mines!"

Lee C. Malone, General Manager of the Fairmont Coal
Company, knew his men by name, for he too had been a miner
and had worked his way up through the ranks. He told a
reporter on the scene that 478 men had been checked off as they
entered the mines that morning. This figure, he said, did not
include the approximately one hundred trappers, mule drivers,
pumpers, and other men not subject to the check system. Later,
Malone drastically revised downward these initial figures.

In the early days of the industry, as we saw in Chapter
One, miners were paid not by the day, but by the amount of coal
they produced during a shift. It was not unusual, therefore, for
fathers to take young sons with them into the pits—for their
small bodies could get into productive places the fathers' could
not. These young men never went through the check system.
Also, individual miners could and did contract with friends
and relatives to dig coal, and share the wages. This "buddy" or
"pal" system was particularly useful when mines were at peak

employment, and new arrivals found it hard to get work.

The early coal mining towns in the United States, especially those within relatively easy traveling distance from the Atlantic coast, attracted hundreds of immigrants who wanted to lose their identity. For what to them were good reasons, they wished no one to know who they were or where they came from. Accordingly, many such immigrants were transients who disdained permanent employment, working a day or two here, and a week or so there, and then drifting on. The "buddy" system, a sort of ultra private enterprise, worked perfectly for them.

That these immigrants succeeded in their quest for anonymity is attested to by one Monongah grave-digger's claim (though not to the Coroner's Jury) that he personally saw six hundred twenty graves prepared in the days immediately following the explosion. Today in Monongah, unmarked graves in "Potter's Field" give further testimony to the veracity of the grave-digger's statement. It seems likely, therefore, that between six hundred and seven hundred men died on that dreadful "Black Friday" in Monongah, West Virginia. Officially, however, the casualty list was finally fixed at three hundred sixty one. Even with this drastically reduced number, the explosion at Monongah remains the worst coal-mine accident in the history of the United States.

Rescue operations, conducted by volunteers from Monongah, Fairmont, and Clarksburg, began just two hours after the accident. The men entered Six almost immediately, while access to Eight was delayed several hours because of wrecked debris blocking the pit-mouth.

At the bottom of Six slope, the wrecked trip was found, with cars (only a portion of the six hundred cars in the two mines) "smashed and piled in all shapes." At the foot of the slope, in a small shanty, rescue workers found the bodies of the slope-tender, a motorman, and a brakeman. One had been eating, and his features had not been changed by death. It was as though a sudden, all-embracing pressure had hit him, suspending him in time and space. The other victims at this site showed no sign of violence—no evidence of any attempted escape. Death must have come to them very unexpectedly and swiftly.

On into the afternoon the grim work continued. As the men worked their way along the heading away from the wrecked trip toward the connecting point with Eight, they had to claw through rock falls and other debris of all kinds, and risk additional explosions. Restoration of full ventilation by late afternoon aided their efforts considerably. Finally, nearing the connection with Eight, the rescue party discovered another wrecked motor and trip, so massive that it would take hours or days to clear up. And they didn't have hours or days—there was always the possibility, the hope, of survivors.

At this wreck, approximately 2,500 feet from the mishap at Six slope, the motorman had been blown to bits, with pieces of his body embedded into the machinery. The brakeman was found a short distance away, his arms, legs, and head severed from his body. His watch, though, in the pocket of his vest—hanging on a peg a few feet uptrack—was ticking merrily along. The difference in the conditions of the bodies suggests that an explosion destroyed Eight, while an implosion or sudden, awful pressure, killed the miners in Six.

Number Eight was the big problem. Entrance to it was blocked from below by the wrecked motor and trip in Six. And from above? The explosion, said the *Fairmont Times*, tore off a concrete roof of the engine house. A large part of the fan was embedded in the mud on the other side of the river, and its fragments were scattered as far as a half mile away. The iron end-gate of a mining car was also blown across the West Fork and stuck into the bank opposite the mines. One witness stated that flames shot one hundred feet out of Eight. A further description was rendered by the *Dispatch*:

> At the entry of Number Eight Mine, the mountainside has been hollowed for a distance of fully a hundred feet. Hundreds of tons of earth and rock were carried away, and it was not until long past midnight that a fan was rigged at this point, and it was possible for the rescuers to make an entry. They were able to reach only two hundred feet when they were stopped by an impassable barrier.... A huge ten ton motor had been completely overturned, barring further progress, and beyond it were tons of earth and wreckage of the cars.

In the darkest hours, however, one should always seek a ray of light. And one came to that saddened village about four that afternoon. Near the entrance to Number Eight, in the area just before the mountain makes its precipitous climb, were a few natural depressions in the earth, known as "sinkholes," or "toadholes." Mothers always warned their children to stay away from them for they could, and often did, fall through into the Number Eight headings.

The authorities ordered that guards be posted at these toadholes to keep zealous friends and bereaved relatives from going in after their loved ones. Late in the day the congregants at a toadhole were startled by loud moans coming from below. Immediately, a volunteer put a heavy cloth around his face, and was lowered by a rope into the hole. Surveying the scene about him, the rescuer saw a dazed, stunned man sitting a hundred feet downtrack. Swiftly, he got to the miner, tied a rope about him, and led him toward the sunshine. All the while, the miner chattered wildly in Polish, a language that the rescuer could not understand. The Pole, Peter Urban by name, was the only survivor of the holocaust. His brother, Stanislaus, was with him at the toadhole, but he could not be revived.

Reports of the day vary considerably about survivors of the Monongah horror, and of how they were rescued. For example, the *Dispatch* stated that the sole survivor was named Peter Roisberg (which could very well have been Urban, for many miners used different names for payroll purposes), and that he was rescued by a group of men working their way through Number Eight. The reporter sensationalized: "A short way down, they [the rescuers] met a sight that chilled their blood. Peter Roisberg, a Polish miner, sat upon the corpse of his brother. Roisberg's mind was in another world. He endeavored to drive off the rescuers and clung to the charred form [presumably of his brother] which had been his seat. It required the efforts of all the men to hold the demented man. . . ."

Back at the Urban residence, Peter's wife Caroline "yelled and hollered because I knew what had happened when the house started shaking. The bodies were so bad that I only recognized one man, because he always wore an earring." She wept through the morning, bitterly lamenting the decision to come to the United States. Like so many other immigrants, she and Peter—a timberman in Romania—had heard just the positive side of America. If they had known that only underground work awaited this proud man of the forest, they would not have immigrated.

Twelve hours after the explosion, Caroline heard a report that there had been one survivor.

Photo of Peter Urban, sole survivor of the Monongah explosion; December 6, 1907. He was later killed by a rock-fall in Number Eight, "his mine." Photo courtesy, Stan Urban (Peter's son); Monongah, West Virginia.

"What was his tag number?" she forlornly asked.

"Fifty-three," replied an official.

Caroline wept anew, but now for joy, because tag-number fifty-three belonged to Peter Urban.

Peter, of course, could remember little about the ordeal: "I know nothing that happened. We went to work that morning and...digged [sic] coal. Then we went and ate...While we were eating there was noise and a report; I told my brother [Stanislaus] we had better run, for something had happened. We did not start at once as my brother thought at first that nothing serious had happened. Soon it became so hot that we started to run. I do not know how I got out of the mine. When I came to, I was at home."

Peter convalesced for three months, and then went back to work in Number Eight, "his mine." Though he had been an outdoorsman in the old country, apparently the lure of the "black gold" got to him and, as with so many other miners, made him love the thing he hated. Soon after returning to his job, Peter's leg was broken in a minor accident; then his back was slightly injured; and in 1926, a sizeable portion of the Number Eight room where he was working caved in, causing a small number of fatalities. Peter Urban was one of them. Caroline, who celebrated her one hundredth birthday in December, 1977, still shook her head at the irony of it all.

There were numerous rumors around Monongah and Fairmont that other men besides Urban escaped. The *Fairmont Times*, December 8, 1907, ran a story about the escape of Dan and Joe Depatris, and Lloyd and Daniel Domenico from Number Eight. The report stated that the quartet were at three thousand feet when the explosion occurred, and that they made it to a toadhole. This seems unlikely, for at three thousand feet, it would take even a physically capable person several minutes to reach a toadhole, let alone someone who had just been stunned by an explosion. Moreover, nothing more is heard of any survivors named Depatris and Domenico. If they did survive, they left the area quickly and without notice. There was a Depatris on the casualty list, but his first name was Felice. There were no Domenicos on the list. It seems unquestionably true that the sole survivor of the Monongah explosion was Peter Urban.

The rescue party continued to inch its way down Number Eight. Men avidly went to work trying to get past the wrecked motor, and with their picks and shovels they attempted to move the dirt and debris that blocked their path. Several men collapsed from the effects of the afterdamp and had to be taken from the scene and treated at field hospitals set up outside on the mine grounds. Later, two of the rescue workers died from complications that resulted from exposure to this deadly essence. Thus, the official casualty list was raised to three hundred sixty-one. All rescue work had to end until ventilation systems sufficient to draw off the afterdamp were installed. It was not until around midnight that the men were allowed to go back into the mine.

Just before the rescuers were to resume their work, a photographer, Marvin D. Boland, slipped unnoticed through a toadhole into Number Eight. He wanted to get dramatic, exclusive pictures; therefore, he brought with him goodly quantities of photo powder. He took his first shot, causing a large flash of light and a report which was audible some distance away. Everyone thought that another explosion had occurred. Then Boland was discovered and rather unceremoniously led off the premises. Some of the spectators were amused by the incident, but the officials were not.

And now they encountered yet another trial. It was as though all the gods of misery and damnation had gathered here in this little hamlet of Monongah, West Virginia. Fire of an undetermined origin flashed through the headings and chambers about four hundred fifty feet from the pit mouth. From one to five a.m., the rescue workers and citizen volunteers formed bucket brigades, taking water from the West Fork, finally—finally—extinguishing the blaze. As the sun rose on Saturday, December 7, 1907, it found an all but exhausted community of men, women, and children. But there was much yet to be done; no time to rest. At five a.m., rescue work was recommenced.

By now, Saturday morning, the labor of salvation was resolved into a systematic operation under the command of James W. Paul, Chief Mine Inspector for West Virginia. All twelve district inspectors were also on the scene, six assigned to each mine, taking turns advising the rescuers and leading search parties. Each shift, which employed forty to fifty men, was organized into crews composed of searchers, bratticemen, and stretcher gangs, all under the supervision of one man.

The crews were further divided into squads of varying

responsibilities. There was a Materials Squad, with the job of supplying lumber and posts to shore up weak places in the mines. There was a Food and Drink Squad. The members of the Messenger Squad ran many miles a shift with notes to and from the inside and outside supervisors. A typical message was the pencilled one sent from John G. Smyth inside the mine to C.E. Scott, outside: "All O.K. here. Now are on the way down to air course, and will be in sixth butt in a short time. Have some bodies in sight, and a horse to fix up." Though telephones were in use in 1907, apparently it was not possible or feasible to set up field phones. Thus, the time-tested procedure of foot-power was utilized.

Once inside the mines, the crews (all members of which had been issued battery operated Wolfe safety lamps) worked according to a pre-arranged plan. Frank Haas, Assistant General Manager for Fairmont, described the methods:

> The various crews under their respective foremen would brattice...clear up falls, and make ready...for exploration, after which the leaders would explore the rooms and working places for bodies.... As soon as a body was located, it was disinfected with a solution of carbolic acid, prescribed by physicians in charge and properly marked with all available information to aid in identification. The party was organized into stretcher crews, and certain men selected to handle the bodies. These, wearing rubber gloves, carefully wrapped the bodies in sheeting and placed them on stretchers which were carried to a truck on the nearest available track. They were then hauled to the bottom of the slope by horses. Notice was given to those in charge outside and when ambulances were ready, the bodies were hoisted to the surface, and taken to the morgue.

And did those people have a doleful job to do! All the bodies they found in the headings, or haul-ways, had been blown to pieces. One miner had been thrown with such force and violence that his body was "mashed flat and left sticking to the coal." Another miner's body was taken from a heading in Eight in no less than twenty pieces. His face was missing, and he wore no clothing. It took authorities several days to establish his identity. Yet another miner, a driver, was covered by his loaded trip, one foot sticking out from the tons of coal on top of him. His horse, completely stripped of harness, was one hundred yards uptrack, lodged solidly in a fall of rock. Another horse was scattered for five hundred feet, torn so badly that its remains, after being sprinkled with carbolic acid and slaked lime, had to be scooped up with shovels; near-by, however, in a small hole off the heading, the horse's master, a trapper boy, was in a sitting position without a mark or scratch anywhere on him.

Away from the headings, in the rooms, bodies were found in many different conditions and positions. Some were stilled in the act of bringing back a pick to dig out coal; others had perhaps been taking a work-break for they were sitting, holding their heads in their hands. Or, perhaps, they had been praying. In many of the rooms, watches left in coats were still ticking, and "picks and shovels were found leaning against props and cars where they had been left."

The greatly varying conditions of the victims once again suggest that the explosion raced along the headings, mutilating everything in its path; and that its force caused an enormous heat and pressure in all the rooms. Surely this explains why some bodies were torn apart, while others remained in a perfectly preserved state.

In addition to coping with the stinging eyes, headaches, nausea, and prostration caused by afterdamp and foul air, the rescuers also encountered intense heat and stench from quickly decaying bodies of humans and horses. Also, in places, whole trips of coal cars, both loaded and empty, blocked their passage. Precious time had to be used to get through these obstacles. If not wrecked trips and fallen debris, it was often the carcass of one of the thirty or so horses and mules taken into Eight on Friday morning. The men resorted to block and tackle as the most proficient method of removing the animals. Sometimes, though, as the beast skidded along the mine floor, it hit a fall or a wrecked motor, thus stopping the efficiency of the block and tackle. This situation necessitated the formation of another squad, which had the gruesome task of sawing off legs and heads to make movement possible.

Little wonder then, indeed, that numerous rescue workers were overcome. They suffered the trauma of witnessing horrifying spectacles, and many of them were affected by this for the rest of their lives. Certainly, all bore the marks of exhaustion, but many came down with respiratory problems caused, undoubtedly, by the afterdamp. Pneumonia, the most fearful complication, hit many of these brave men, producing yet another problem for the beleaguered community.

Outside, meanwhile, horrendous scenes unfolded. All through the day, terror stricken men, women, children, and old miners hovered at the mines' entrances, many of them trying to get in. One strong Polish woman did break through the barrier and head frenziedly into Number Eight. It took several men to subdue her. Another woman tried to fling herself into the West Fork, to end her misery. All day long the air was splintered with shrieks of horror, and also with wild laughter that bordered on insanity. Constantly, the women wanted the officials to tell them the fate of their husbands, fathers, and sons. The officials, of course, could do nothing at this point, causing some women to tear their hair from their heads in great handfuls, and to disfigure their pretty faces by plunging sharp fingernails into them. One young wife, it was reported, lay down on the frozen ground and cried herself to sleep. In that condition, she was taken to her house near-by. When two bodies were brought out during the early period of rescue operations, a riot erupted among those who wanted to identify them.

Waiting for news after Monongah explosion in West Virginia; December 6, 1907

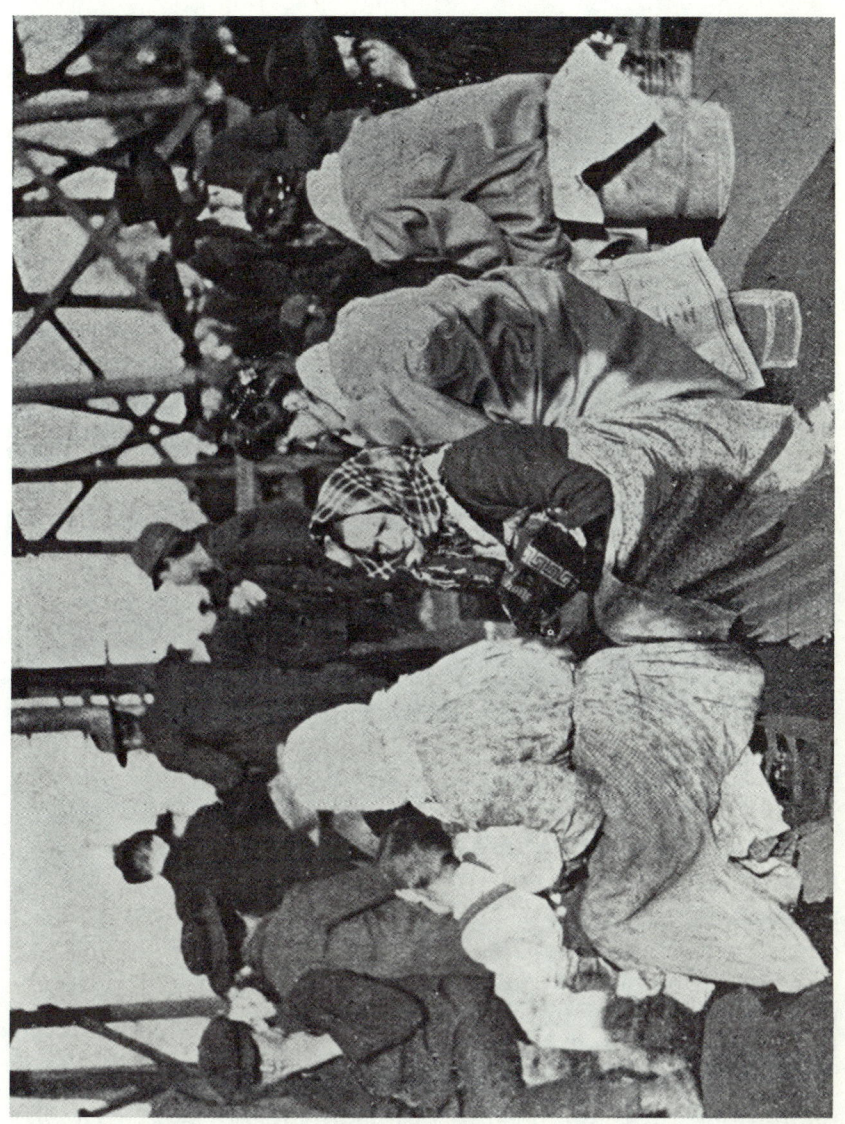

Monongah, December 6, 1907. Photo courtesy, UMWA.

Crowds viewing bodies after Monongah explosion; December 6, 1907.

Finally, it was necessary for nearly every able-bodied man in Monongah, Fairmont, and Clarksburg, to be mobilized to control the mourners. West Virginia Governor W.M.O. Dawson was criticized in some circles for not mobilizing the National Guard. He did suggest its use, however, to control the crowds. Fairmont Coal Company President Watson refused the offer, citing the general orderliness of the crowd. The Governor sent Colonel Joe McDermott to Monongah as his personal representative. Clarence Hall, a federal government explosives expert, arrived as the official envoy of President Roosevelt.

The problem was compounded by the inability of most of the miners' families to speak English. The majority were from Italy, and the others from Poland*, Turkey, Austria and Russia. Even if they were inclined to observe needed rules and regulations, the language barrier in many instances kept them from doing it.

*There was no Poland, as such, in 1907. The ancient territories of that country had been absorbed by Germany, Russia, and Austria. Poland was resurrected as an independent country at the end of World War I. It was possible in 1907, therefore, for a "Pole"—at least, according to some Americans—to come from Romania, Lithuania, Hungary, Russia, or even Turkey.

Besides the mourners there were, naturally, hundreds of the morbidly curious who came to see the sights at Monongah. A large proportion of the area's working force was on a week-end holiday, so they came by buckboard, by horse, by trolley car (The Fairmont Traction Company, doing a boom business, put several "special cars" into operation), and by foot.

A rumor had got out that blood was gushing from the pit-mouth of Number Eight, and who could resist an opportunity to witness that? Also, it was reported that the body of a huge black cat had been found in Number Six. It had only one eye, so it was said, right in the center of its face. Dr. H.S. Keister of Fairmont was supposed to have quoted the examining physician (without naming him) that the cat had never had but one eye; there was no place for another eye to be.** Nobody, however, could find this monster's body.

The crowd, though large (some estimates say 10,000; others, 25,000) was mostly orderly, surging forward only when it thought there was something macabre to see. Scuffles broke out from time to time among spectators for the most favorable vantage points. Many of the visitors were dressed in their Sunday finery because it was the week-end. By late Saturday, people were at the mines from virtually every city of West Virginia, and from most parts of Maryland and Pennsylvania.

**Stories about the "black cat" abound in the northern parts of West Virginia. For example, in Murray, West Virginia, little three year old Frank Sprouse was run over and killed by a train. Both before and after, a large, black, one-eyed cat walked and ran around the boy. Thomas Branson of Fairmont, a brickmason, was seen one day with a big cat walking alongside him. A few days later, the building on which Branson was working collapsed, killing the brickmason. Another story relates that in Grafton, sometime in 1904, a man named Sariti walked down a street leading a black, one-eyed cat on a leash. One day Sariti boarded a train for Clarksburg, but had told a lady friend, with no explanation, that he really was going to Fayette to commit suicide. The woman thought nothing more about Sariti's claim until she saw a newspaper heading: "Man fined for taking his own life." The story's dateline was Fayette, West Virginia, and the victim, sure enough, was Mr. Sariti, who had jumped into the New River. Mayor Shaffer of Fayette was apparently fed up with people using his river for such purposes; thus, when $130.00 were found in Sariti's pockets, the mayor fined him $80.50. Sariti's father, from Pittsburgh, tried to recover the cash, but was himself fined $17.00, again by the enterprising mayor. According to reports, Sariti senior brought the big cat back to Grafton and gave it to a lady who lived on Latrobe Street. It supposedly continued its work of mischief after its sojourn in Fayette. Interestingly enough, however, nothing more is heard of it after December 6, 1907.

A special excursion train was chartered in Baltimore for a Monongah outing.

Miraculously, vandalism was not a great problem during the crisis, even with the huge gatherings. In part this may have been due to no drinking diversions, since all saloons were closed within an hour after the explosion. Only one house was reported to have been ransacked, in West Monongah. Five miners, all victims, had lived there. Everything of value had been taken from the dwelling, and all the mattresses and beds had been torn apart in search of precious possessions. Perhaps the thieves were suddenly stricken with the utter outrageousness of their deeds, for no other miners' houses were robbed.

At four in the afternoon Sunday, December 8, the mobs were dispersed, and for a very good reason. The authorities did not have to coax too many to leave, for a new fire had broken out in Number Eight. It was down about 1,500 feet, and before long flames began to leap through several toadholes. There was a real threat of another explosion. The *Dispatch* reporter gave this description: "...The guards announced that another explosion was liable to happen at any minute. Instantly, the crowd scattered along the trolley-tracks, and over the fields, and across the bridge, and some ran pell-mell into town. There were some in the throng about the mine entrances, however, who did not join in the mine panic. These were men and women who had dear ones in the smoking entries."

The fresh blaze brought all rescue work to a halt. The men were ordered to come immediately to the surface. In a few hours, however, the fire was put out, and work was resumed by early evening.

By now, of course, many bodies had been recovered and, after a time of confusion, an organized procedure for processing them was developed. A workroom on the mine grounds was transformed into a temporary morgue, where three undertakers and thirty embalmers worked in relays of about eight hours. Each body was placed on a worktable and a numbered card attached to it. A record of each body was kept by the coroner's office, and by the Fairmont Coal Company. One of the two payroll clerks at the mines was always present to help with identification. Some miners were so badly mutilated, however, that not even the payroll clerks could say very quickly who they were.

As the clothes were removed from the dead men, valuables and money were collected, and then turned over to the coroner's office. In due time he, Coroner E.S. Amos, deposited $4,367.12 in a Fairmont bank, except $23.19 found on the body of Andrew Morris, and turned over to his widow. No reference was made to why (except that many bodies were un-identifiable) it was apparently not possible for the larger figure also to have been returned to family members. There are no references, either, to the ultimate disposition of this money. Presumably, it became a part of the relief fund started by the citizens of Monogah and Fairmont. One miner, it was known, an Italian, had $2,500 on his person when he entered the mine Friday morning. The dollars, all in a money belt, were never recovered—at least by those who should have received them.

By late Friday the supply of coffins in Monongah and Fairmont was exhausted. Priority orders went through to Pittsburgh and Zanesville, Ohio, so that by Saturday afternoon, great stacks of burial boxes lined the streets of Monongah. Each coffin was a plain rectangular box with no inside lining. Several people were hired to tack black lacing in the casket, so that the occupant would not be facing bare, wooden walls. Some of the dead miners' relatives objected to this procedure, and as a result in a few instances, the coal company procured suitable burial containers.

Very quickly the victims' bodies were processed and placed in coffins. The Americans were sent to their homes. The Italians, however, were dispatched directly to the Monongah cemetery. A morgue had been set up there, consisting of tents furnished by Captain M.M. Neely of the National Guard. The Slavonic victims were taken to the Polish Catholic Church. These procedures, said the officials, prevented congestion, though some argued that they reflected native prejudices.

A funeral wagon taking a body from the morgue to a Polish cemetery at Thoburn—a little village adjoining Monongah— overturned on a steep hill. The casket inside was not disturbed, but the driver suffered a sprained back. The horses broke loose from the wagon and ran pell-mell through the area. They crashed into J.A. Clark's buggy, causing almost total damage. Finally, the maddened beasts fell headlong into the West Fork where they roiled about for several minutes. Then, somehow regaining their composure, they swam to the bank and were re-harnessed by a group of waiting men.

Coffins for Monongah victims; December, 1907. Photo courtesy, UMWA.

The body of John Hearmans, an American, was placed in the front parlor of the cottage where he had lived with his wife and four children. Mrs. Hearmans at this very instant, was giving birth upstairs to their fifth child. The *Fairmont Times* reported: "Mrs. Hearmans' condition is very serious. She is aware of the fact that there has been an explosion. She realizes that the entombed miners are dead beyond peradventure, and knows that the little son whose coming has been looked forward to for months is born in the world an orphan."

While Mrs. Hearmans at least had the comfort of a decent burial for her husband, several other widows never did experience the recovery of the lost men from the mines. One such was Mrs. G.L. Davis. After the explosion, for several years, Mrs. Davis went down the hill every day the mines operated, and gathered a burlap sack full of coal from the cars that exited from the pits. She became a familiar sight to the area's residents as she carried the coal up the mountain and deposited it in a pile near her house. She never burned any of it, or allowed anyone else to. As Lacey Dillon, in *They Died in Darkness* reports, "When asked why she piled this unused coal daily, she stated that she had hopes of retrieving some of her lost husband's body. She was a young woman when the tragedy happened, and she lived to be an old lady. At her death a few years ago, her sons gave the coal to the churches of Monongah. The coal pile had grown to an enormous size."

That Sabbath, December 8, 1907, in the Fairmont region was one to be remembered. Rescue work was continuing apace at Numbers Six and Eight, bodies were being identified and processed with ever accelerating rapidity at the morgue—now situated in the new First National Bank Building—and sent to various places to await burial. And the crowds continued to grow.

Most of the victims at the mine explosion were Italian—one hundred seventy-one. Father Joseph D'Andrea of Our Lady of Pompeii Roman Catholic Church was, therefore, overtaxed in his efforts to comfort the bereaved, to provide extreme unction to the departed, and to supervise the burials. It was an awesome, heart-breaking task. At the Father's services that Sunday, the *Dispatch* reporter stated that "women, children, and even strong hearted men were overcome by their emotions and at least a dozen had to be assisted to the open air where they were revived and led to their home." Father

D'Andrea estimated that over forty of his families had been bereft of their sole support. This was, unfortunately, a conservative estimate.

No less severe were the duties and responsibilities of Father John Lekstrom of St. Stanislaus Church, where the predominant membership of Slavonic miners and their families had always worshiped. Nearly a hundred of his parishoners lay dead on this day, and it took considerable effort on his part to keep from being himself overcome by grief. A requiem mass was scheduled at 10:30 that morning, but at least an hour in advance, scores of women congregated outside the church. They were all clad in the weeds of widowhood, accompanied by little children who did not quite understand all these fearful events swirling about them. All of them, women, children, and men conversed in subdued tones. Then they saw the tall, lean figure of Father Lekstrom coming through the fields toward his little church. At his appearance, the parishoners could no longer contain their grief and they broke into convulsive weeping.

"Oh my poor people!" Father Lekstrom exclaimed, as he gazed affectionately at his flock. Then he led them into the little red church. At the threshold, many knelt and kissed the sacred step, made the sign of the cross, and then sadly took their seats.

As Father Lekstrom entered the church, he told reporters: "I shall refrain from mentioning the disaster.... To recall the horror of all this would result in the women losing their composure, and I know I should become similarly affected. Moreover, this is a time of deeds and mercy, not idle words of commiseration. We are doing all in our power to provide for the distressed of our parish." As the services got under way, Father Lekstrom was true to his word: he did not mention the tragedy. But only a few hundred feet away, at the mines, the loud work of reconstructing the fan house at Eight kept the worshipers reminded, though such was unnecessary, that reality was close at hand.

The Protestant denominations were stunned too, and the pastors of the larger churches—the Reverends J.C. Bloomfield, Methodist-Protestant Temple; H.G. Stoetzer, First Presbyterian of Fairmont; and W.G. David, Presbyterian of Monongah—kept their services short. It was impossible for them to give extended eulogies because of fainting women, weeping children, and trembling men. The consensus of religious opinion seemed to be that the disaster was a visitation from an all wise providence, and that the survivors could glean many valuable lessons from it.

At the services a week later, the Reverend Bloomfield tried, more or less successfully, to cast a new light upon the tragedy, hoping to relieve God of some of the blame that had been heaped on Him for the explosion. He worked feverishly the entire week on the sermon. When it was finished, he titled it, "The Bright Side of the Monongah Disaster." He took his text from Luke 13: 4 and 5:

> Or those eighteen, upon whom the tower in Siloam fell,
> and slew them, think ye that they were sinners above all men that dwelt in Jerusalem?
> I tell you, Nay; but except ye repent, ye shall all likewise perish.

and from I Kings 19: 12:

> And after the earthquake a fire: *but* the Lord *was* not in the fire; and after a fire a still small voice.

These chapters and verses, according to Bloomfield, showed the relationship between sin and suffering. Properly thought out, the Monongah disaster could have the positive influence of causing people to do right in this world. He said:

> The Monongah disaster thrust itself upon us as a problem, and we are conscious of attempts to solve it. The solution is sought so as to locate the responsibility. That there is responsibility, no-one denies, but where and on whom no-one knows. Many people will refer to the disaster in solemn tones as an awful providence of God. Why not say, 'what an awful consequence of human blundering?' Many theories have been advanced as possible explanations of the explosion. Nought of them has human blunders as their basis. For example, if we accept the theory of the broken trip, then unfaithfulness in the weaving of the cables, or in the forging of the coupling, or carelessness in overtaxing the cables or couplings was responsible for this

calamity. Or if a blown out shot igniting the gas and dust is accepted as the explanation for the explosion, human blundering is self evident. After all the theories have been advanced, we must admit that the responsibilities for Monongah rest on human shoulders. Man violated God's ordained law.

The minister then turned his attention to the distinct possibility that the mine explosion could renew not only man's faith in God, but also man's faith in himself:

The disaster has increased the faith in our fellow man. Our common grief has buried social and commercial distinctions. [This remark was occasioned, apparently, by the other coal companies in the area making no effort to take advantage of the Fairmont operations, although it is true that all the Fairmont mines were open with, of course, the exceptions of Six and Eight.] The greatest loss any man can sustain is loss of faith in God, and next to that, his loss of faith in man. The seemingly intense selfishness of our age, born of unprecedented competition along all the lines, has a tendency to force men to lose their faith in each other, and any experience, however bad, that will reinforce men's faith in men, is worth the while. Gethsemane and Calvary awaken men's faith in God. Such disasters as this explosion, because of what they bring to the surface, buttress men's faith in each other. Increased sympathy exists between this community and the Fairmont Coal Company, in these days when legal prosecutions of corporations savor somewhat of persecution—you often find communities knocking their leading corporations—and often [we]...have not always been kindly in...criticisms of the coal company. On the other hand, the men of this leading business concern may not always have been as considerate of others as they might have been.

He closed his remarks by asserting that if the horrible events at Monongah caused employers and employees to treat each other with greater fairness and respect than in the past, all would not have been in vain.

The Reverend Stoetzer's sermon on the 15th carried a similar message:

From coal and tar are derived numerous products—beautiful dyes, helpful drugs, sweet perfumes, valuable antiseptics, rare flavors. Out of the great

disaster has come undreamed of sympathy and
generosity; unsuspecting loyalty of employee to
employer, and the greatest imaginable devotion of
the employer to employee. The calamity has driven
avarice and selfishness into darkness, and has
opened foundations of sympathy and springs of
generosity, and has touched slumbering bravery and
daring heroism.

It was brought out also in this and other services that
Friday, December 6, was a "double hoo-doo" day for
Consolidation Coal Corporation, of which Fairmont was a
constituent. Its steamer, the *Thomas W. Lawson,* went down
that day in the stormy waters off the coast of England. The
Monongah-Fairmont ministry, therefore, gave many
expressions of sympathy and concern for the sufferings of the
coal company itself. These were expressions that grew in the
days ahead, so much so that some citizens accused the
churches of caring too much for the company and not enough
for the individual.

In the matter of burial, some officials thought at first about
digging trenches and resorting to mass graves for the
Monongah victims. This suggestion met with such violent
resistance from all quarters that it was soon discarded. In the
cemeteries, therefore, grave rows were dug three feet apart. The
graves were six feet, six inches long; anywhere from four to six
feet deep, and three feet, eight inches wide. The *Fairmont
Times,* December 9, described the grave-digging procedures in
the Italian and Polish cemeteries: "The men work in shifts, the
shovelers following after the digger. In succession, the digger
goes over one grave digging as deep as he can.... Then the
man following shovels out what the digger has just left. This
continues until the men have worked round the cemetery,
where they begin at the start again. Dozens of graves are
finished and dozens of men are buried, and still the gruesome
task goes on."

The good weather ended late Saturday, and now a cold,
drizzly rain oozed down from the heavens, making the earth
soggy and heavy. Several hired grave-diggers fell to with picks
and shovels, but their job was well nigh impossible. A grave
completed one night was almost always found the next
morning half full of water, and had to be bailed out. It was
difficult to keep the men going in the continual downpour. The

Ladies' Aid Society of Fairmont-Monongah furnished them with sandwiches and hot coffee—and so they endured.

The matter of burial also brought out several religious biases and prejudices. Frank Haas gave this description:

> At the time of the explosion, both the Italian and Polish Catholic churches had cemeteries immediately adjoining, separated by a wire fence. At the very start, the men at work in these cemeteries were admonished by the representatives of these two churches to be very careful not to allow any member of the Italian church to be buried in the Polish side, and *vice-versa,* and again later not to allow Protestants to be buried in either of these cemeteries. For these reasons, a new cemetery was located, adjoining the Polish, to be used as a burying ground for Protestants and unknowns. This fact made it necessary to have representatives of the Catholic churches present who had a list of members of their congregation; and whose advice was followed in determining the cemetery in which each body was interred.

Several of the survivors, apparently trying to escape funeral expenses (though there were none), allowed their dead ones to be buried as unknowns. Every morning, for example, sticks and marks were found beside newly made graves. The workmen destroyed these signs each day, only to have them reappear by the next morning. After several days, the signs disappeared for good, probably because the location of the graves became firmly implanted in the sufferers' minds.

Despite hardships, tragedy, and sadness, life goes on though the sufferers might will it otherwise. Thus, the week passed in Monongah, and then the month—but the memories lingered on. Lives were re-fashioned all the way from several foreign women sadly returning to their homes in Europe to numerous widows re-marrying. One Italian woman, on the day after the explosion, demanded a certificate from the coal company showing that her husband was dead. She wanted to marry again, but could not until she had legally been declared a widow. The company complied with her request.

These were the living who for the remainder of their lives would bear the marks of "Black Friday" in Monongah. One of them, unknown, wrote a poem to express the feelings of the day, because very frequently grief will lead one to poetry. The poem was entitled, "The Hero of the Blackened Face." It went this way:

> With blackened face and dusty clothes
> the stalwart miner onward goes.
> With fearless eye, with heaving chest
> He stands apart from all the rest.
> Daily he faces death most drear
> with smiling face and words of cheer.
> Daily he goes his fate to meet
> Hopes to return, loved ones to greet.
> A hero in his manhood's pride, a work he does
> Yet few may know the dangers he undergoes.

> Yet when at last he meets his fate
> And when he stands at Heaven's gate
> He will hear words: 'Well done thou true;
> Enter the eternal brotherhood, thou hero of
> the blackened face.'

What the poem lacked in sophistication, it made up in feeling and timeliness. It hit the citizens of Monongah and Fairmont, and indeed the entire country, with thoughts that would never—could never—be forgotten.

Sources

A While Ago Times, June-July, 1971.

Birmingham *News,* December 6, 1907.

Briggs, Everett F., The Rev. *Catholic Churches of Monongah, West Virginia, 1957. 50th Anniversary Year, Including the Monongah Disaster of 1907.* Pamphlet, Private Printing.

Dillon, Lacey A. *They Died in Darkness.* Ravencliff, West Virginia: McClean Printing Company, undated.

Fairmont *Free Press,* December 12, 1907.

Fairmont *Times,* December 7, 8, 9, 1907.

Fairmont *West Virginian,* December 6, 9, 1907.

Haas, Frank. *The Explosion at Monongah Mines.* Bulletin No. 11, Fairmont Coal Company, 1910.

Interview: Caroline Urban (Peter's wife) and Stanislaus Urban (Peter's son), December 23, 1977.
MSS. West Virginia University.
Pittsburgh *Dispatch*, December 7, 9, 15, 1907.
Pittsburgh *Press,* December 11, 1907.
The *Times-West Virginian,* December 9, 1977.

Chapter Three

The MRC:
A Lesson in Private Welfare

Numbers Six and Eight were two of forty Watson mines in the Fairmont area. Altogether, the Fairmont Coal Company, a subsidiary of Baltimore based Consolidation—also owned by the Watsons—had more than one hundred producing properties in West Virginia and Kentucky, with an annual capacity of six million tons. It owned enough undeveloped land to double its operations within a few years. In 1907, it was capitalized at approximately twenty million dollars.

It was apparently not capable, however, of providing permanently for the 196 widows and 468 children left by the Monongah explosion. This impression was encouraged by the very first appeal letter from Monongah that "the burden...is more than this company or any other company could bear for any length of time." Governor Dawson added to the image in his general appeal of December 14: "The company has generously declared that the families occupying their houses may remain until other provision is made for them, but the operations cannot be resumed at the damaged mines until the houses are available for the new force. The magnitude of this disaster is too great for West Virginia to alone render all the help required. It commands national attention."

It is true that the Company paid funeral expenses, allowed survivors to live in its houses for a limited time, and contributed $17,500 to a general relief fund. In addition to all this, it ultimately gave $150.00 to each widow as a final settlement, and $75.00 to each child under sixteen. Considering that, all told, there were about 1,000 survivors (widows, children, aged parents, and other dependents), the coal company's average payment was $150.00 per family. This was a piddling amount, particularly if any attention was given to the lost earning potential of each of the victims. For that day and time, however, this was indeed a generous response.

The settlement reflected the widespread "industrial feudalism" in the United States, a condition by which the workers were economically bound to the company. They had to pay high rents for company houses, and were sometimes forced to buy their provisions from company commissaries. It was not unknown, as in the case of the Pullman organization in Chicago, for workmen to be paid in scrip which could be negotiated only at a company establishment. The public tended to accept these practices, and local and state politicians always lauded the industry, for it gave jobs to their constituents. Thus, it was not unusual for a company to dictate the economic and social life of an entire community. This is why the outpourings of sympathy after the Monongah disaster seemed to be as much for the Company as for the victims, causing one Fairmont *Times* reporter to give this rhetorical gush: "Whatever criticisms have been made about the safety of West Virginia mines cannot be applied to Monongah. They are the very best in the world." Tell that to the widows and fatherless children of Monongah!

There were no provisions in 1907 for the federal government to participate in welfare schemes for survivors of disasters. Many reformers, especially the so-called "Social Gospelers," urged the passage of laws to protect the laborer and to provide for his family when he could not work. Their plea did get a sympathetic response from President Roosevelt. The thing to overcome, though, was a Congress that viewed any reform of capitalism as "socialistic," and a general public that agreed with that philosophy. It was many years before public and congressional attitudes changed to benefit the common man in the United States.

It fell upon the charitable impulses of the nation's private citizens to give relief to the Monongah sufferers. The day after the explosion, several newspapers, including the *Dispatch* and the Fairmont *Times,* started a relief fund, as did numerous business concerns. The Union Relief Association was the first broadly based organization, composed of citizens from all different professions and occupations. It was soon overshadowed, however, by a group known as the Monongah Relief Committee. This organization (MRC) provided an excellent example of how an extensive, specific welfare program could be conducted by the private sector. Chaired by Monongah Mayor, W.H. Moore, the MRC was supposedly independent of the Fairmont Coal Company, but this did not always turn out to be the case.

The MRC's appeals of December 14 and 27 asked for food, clothing, and money. The response around the state and country was quite positive, for Monongah had now caught national attention. In fact, the initial reaction was so good that it ultimately produced negative effects. For one thing, word got out that the Company itself would care for all the dependents, causing many would-be donors—especially the wealthy ones— to refrain from contributing. For another, reports far and wide had it that the MRC was doing such a grand job that it had already reached its goal of $250,000 (the grand total ultimately collected was $154,261.72), thus obviating the need for professional, outside agencies to offer assistance. Therefore, the Carnegie Relief Commission and the National Red Cross gave only minimal contributions—at least in terms of money.

The Red Cross's most significant role was to borrow a social worker from the Pittsburgh branch of the Russell Sage Foundation and send her to Monongah. That person was Margaret F. Byington, a professional with an A.B. from Wellesley and an M.S. from Columbia. She conducted a much needed census of survivors, and bestowed upon the relief work a steady, systematic operation. But she soon discovered that she could not apply the professional touch without encountering several difficulties.

Many local residents objected to "outsiders" taking over the relief operations. Nowhere was this resentment seen more than in the area's Catholic clergy. Apparently, Fathers D'Andrea and Lekstrom believed that relief funds should be given directly to them and then distributed to the victims. In a letter to Mabel T. Boardman of Red Cross National Headquarters, Miss Byington outlined some of her problems with the clergy:

> Father D'Andrea...accompanied me to all the Italian families, and seemed friendly, though because of his poor English, our discussion of the situation was rather limited.
> Father Lekstrom, the Slavic priest, disappeared for nearly a week, without giving the committee [MRC] an accurate list of the names of his people...which resulted in some difficulties about giving out the money. I saw Father Lekstrom soon after my arrival

and he seemed cordial and friendly. I presume that the difficulty has arisen because these priests feel that the entire responsibility for their people should be left with them, and that the money should pass through their hands. It may be that this would have been wiser for the present. The priests have, however, not seemed very anxious to co-operate with the committee, failing to come to committee meetings.

The Red Cross's relationship with the Monongah priests was placid compared to that of the Bishop of West Virginia, the Right Reverend T.J. Donahue, of Wheeling, a member of the MRC. He was ameliorative in this January, 1908 letter to Miss Boardman:

I beg to acknowledge receipt of your favor of Jan. 18th advising me that Miss Byington was at Monongah aiding in the relief work there as a representative of the Red Cross Society. This I knew from several attendances at the meetings of the General Relief Committee...but at the same time I thank you for the courtesy of your communication. It affords me gratification to write you that I shall be happy to do all I can myself and to direct the local pastors to exert themselves...I am using every effort to collect cash subscriptions also.

Despite the conciliatory tone of this letter, Bishop Donahue believed that he and his priests should act as the sole representatives, and even as the attorneys, of the survivors. This role, said Paul U. Kellogg, director of a group called "Charities and the Commons," would disqualify the clergy "to judge whether or not a possible recipient actually falls within the scope of the relief fund."

The Bishop also felt that his clergy was ignored in carrying out certain essential functions. One of Byington's first tasks, for example, was to ascertain individual, specific needs of survivors. She had difficulty doing this, in part, because of the priests' non-cooperation. Bishop Donahue gave his solution to the problem in this crusty letter, January 16, 1908, to MRC's J.M. Jacobs: "I confess that I am at a loss to understand where all the difficulty lies in procuring a correct list of the relatives and dependents of the mine disaster. If they were given to understand that no notice would be taken of them unless they came forward and proved their claims upon the fund, all concerned would soon display the necessary alacrity."

The situation was relieved somewhat by Byington absenting herself for a week, and by key members of the MRC writing tactful letters to the Bishop. It should be noted that Monongah was not unique in the way it treated Red Cross officials. It was often the case that small town residents believed they were more capable of handling disasters—even the ones of great magnitude—than were outside agencies. This belief persists even today, though not as emphatically as in 1907.

The Red Cross also encountered problems when it solicited relief funds from wealthy companies and individuals. William D. Kelly of Philadelphia's Clearfield Bituminous Coal Corporation, sent this terse reply to the Pennsylvania Branch: "According to the published reports of the Company [Fairmont], they are making a great deal of money over and above their fixed charges, and it does not seem to me that it is quite in order for the Red Cross to be called in to contribute when that Company should take care of its own people."

The Red Cross faced yet another obstacle when it tried to deal with John D. Rockefeller, Jr. An exchange of letters between Rockefeller and Boardman gave a good insight into the former's philosophy of giving. In the first letter, December 27, 1907, Rockefeller mentioned that he had received an appeal from the MRC. Showing that he would refuse individual, direct aid, he said, "Since the Committee [MRC] in charge of this work is local, and since, so far as I am able to ascertain, no committee has been organized in New York to cooperate with the local committee, nor has any public effort been made in New York to secure funds for the committee, we are at a loss to know what is needed in the premises and how far the people of this country at large should be called to contribute to this fund."

His second letter, a month later, was much bolder than the first: "It is evident that more harm than good would be done by further contributions at present to the sufferers from these two mining disasters. [The reference to the second disaster was the Darr explosion in Pennsylvania on December 19.] It is a fair question as to how far the charitably inclined people in this country are under obligation to provide for the relatives in foreign countries of the miners who lost their lives. The obligation seems to me very remote." This suggestion was ultimately rejected by the MRC.

In the third letter, February 7, 1908, Rockefeller terminated any relief role he might have played by asserting: "There seems to be quite as much danger that the public will overdo in relief work as that it will not do enough, and I verily believe that quite as much harm is done through the former as through the latter." This remark was occasioned, in part, by continuing reports that the MRC had already reached its quarter-million goal, and that contributions now would be mere gratuities.

The Monongah Relief Committee, meanwhile, went steadily along with its work of appeal, causing an overwhelming response from the entire country. Over two hundred newspapers in the large cities printed the appeal, and before long so much material began to arrive in Monongah that it was difficult to keep track of it. Some citizens pointed out that because of this national relief, the Monongah survivors were now living better than they ever had before. Indeed, some began to argue that an economic imbalance was created by all the charitable activities. Within the mining camps themselves, several quarrels erupted among survivors over who was getting what, and how much. Ironically, the one family directly connected with the explosion who got nothing from the MRC was that of Peter Urban. Though he was unable to work, he was nevertheless a survivor, and therefore not eligible for aid.

Among other things, these came to the MRC:

. Seventy-two pairs of shoes from Hoge-Montgomery Company, Frankfort, Kentucky.

. A car-load of apples from C.L. Robinson of Winchester, Virginia. It was found "impracticable," however, to distribute the apples among the Monongah sufferers. The coal company, therefore, bought them, and added $302.00 to the general relief fund.

. A barrel of "Dewey's Best Flour," from Dewey Brothers, Blanchester, Ohio.

. Twenty-seven cases of "butterine" from the Capital City Dairy Company, Columbus, Ohio

. A box of coats from Spring-Holzworth Company, Alliance, Ohio.

. A carton of "tiger-biscuits" from Ward-Mackey Company, Pittsburgh.

. Free use of several typewriters from the Smith Premier

Typewriter Company, Pittsburgh.

. Two barrels of clothing from the Haverford, Pennsylvania Branch of the Needlework Guild of America.

Other people who were applied to, however, were not so eager to help. One former resident of Monongah had lost $700.00 in a business venture there, and this was sufficient reason for him to ignore the relief fund, and adjure all his friends and aquaintances to do the same thing.

Mrs. John McGann of Sabula, Iowa, wrote, offering to adopt a youngster from the Monongah disaster. She wanted a girl, ten to twelve years old, who was healthy, for "we own a farm of our own and the adopted daughter will fall air [sic] to it one day." Also, N.O. Sowders, Superintendent of the Children's Home Society of West Virginia, sent several forms to the MRC, "for use of those who wish to convey a child to our legal custody. The child must be healthy and sound of mind, and be under twelve years old." No evidence exists to show whether any children of the Monongah disaster were ever adopted.

The catastrophe opened the door of opportunity to several entrepreneurs. The British-American Land and Investment Company, for example, offered to sell acres of "Paradise Farms," at Texarkana, Texas. The victims of the Monongah blast could have a new start in life, and raise collards, onions, spinach, and lettuce, in a place where Christmas Day was like summer. Paradise Farms had been inspected by the U.S. Government's "Soil Survey," and was found to be the "greatest poor man's country in the whole world."

The Company's first letter, signed by J.B. Ladd, went to none other than Governor Dawson. Dated January 2, 1908, it promised: "If you will send these families to Texarkana, and your association can buy tracts for them on a small payment basis, there will be absolutely no question about their future support. We'll make a special offer to you and throw off ten dollars per acre to any of these families that you or your committee recommend, and will take a payment of $5.00 per tract down, and $1.00 per week without interest or taxes until the land is paid for." By doing this, said Mr. Ladd, the company was "donating practically $50.00 to each family that might go into the district."

The Governor did not respond to this; thus, on January 29, Ladd wrote another letter, this time to Joseph Sands of the

MRC. He now stated that the families "could clear from fifty to two hundred dollars a month off the produce." The tracts at Paradise Farms sold for $300.00 each "without any interest or taxes for the first year." Ladd sent along with this letter several testimonials from people already living at Paradise Farms. Sands answered that the MRC had nothing to do with where the victims and their families went. Apparently none of the victims took advantage of this "generous" offer of the British-American Land Development Company.

Another similar offer to "help" came, not from a company but from an individual. Aspiring and enterprising authoress, Mrs. Charlton Edholm from a small town in California wrote this to the Committee: "In response to your circular, I wish to offer you $15.00 worth of my books 'Traffic in Girls.' They retail at 30¢ and sell readily, and no doubt you could find agents willing to sell them without commission, for the good of the cause." The MRC never responded to Mrs. Edholm's offer, and the world had to keep waiting for her *magnum opus*.

By mid-January the MRC's tasks had become systematized, thanks largely to Miss Byington's expertise in organization. A printed form was now used to gather as much information about survivors as possible, and to ascertain their needs. The Committee did become, however, a sort of "moral guardian" in its efforts at charity. On the forms were spaces for "plans of widow," "general condition of dependents," "appearance," "command of English," "has woman kept boarders?" and "note here any further information, desirable impressions, etc." Negative reports (which were more plentiful for the foreign than for the American element) on these forms, however, did not affect a widow or a dependent's eligibility. Apparently, one of the chief functions of the form was to identify the proper people to receive any money that the MRC might ultimately distribute.

The Committee became somewhat paranoic in protecting funds it had collected. It is true that at least six people from other cities, who had no relatives working at the mines, presented claims to the MRC, and that much precious time was spent invalidating them. There were, however, several cases of immediate and legitimate need for which the MRC required extensive documentation. Each time the Committee met, groups of widows petitioned for help. The meeting of May 22, 1908, was typical: Mrs. Floy Ford wanted $150.00 to pay her bills which had been accumulating since December; Edith Masch requested a similar amount for a payment on her house; Beatrice Spragg needed $75.00 for bills; and Mrs. William Walls $105.00 to pay for property. Generally, after each widow proved that her claim was just, her request was usually, but not always, granted. The money given was an advance against the final settlement with the MRC.

The Committee faced another problem when it began to receive unexpected and unwelcome bills from the Jones Granite Company for tombstones. Apparently, several widows had bought these materials and had the bills forwarded to the MRC. Maggie Snodgrass and Annie Hamm each indebted the Committee $95.00; Mrs. Fred Cooper, $80.00; and Mrs. G.L. Davis, $90.00—despite the fact that her husband was not recovered from the mines. The Relief Fund, however, would not see to any of these bills, claiming that if it dealt directly with companies instead of victims, its accounts would be vastly complicated. Later, when each survivor received a settlement from the MRC, presumably Mr. Jones got his money.

As the MRC's work became more widely known, numerous individuals began to write to it on a variety of subjects. Some letters, like the one received about Victoria Susnofsky, were spiteful: "This woman was formerly Sim Schultze's wife. Schultze was killed in No. 5 about 8 months ago by fall of slate. She left and returned soon with above man [Susnofsky] as husband.... She is on 'fast order' drinks, and will promise anything for money." Another letter asked for an investigation into the life-style of Gertie Watkins, who had been induced to live with a family in Grafton, "who wanted to profit with some of the money she receives from these funds." Many letters came in from the Fairmont Coal Company, informing the Committee of wages paid to victims up to the time of the explosion, and how much (or how little) the victim had supported his dependents. The relationship between the MRC and the coal company grew ever closer, and within a few weeks of organizing, the Committee began to use the Company's private telegraph line to make appeals, and to get information. By late January, 1908, the Committee had reached a tentative conclusion on how to distribute the collected funds. Some widows could be pensioned at two to four dollars a week, if they chose, until a final compensation was decided. Others, mostly foreigners, could take a lump sum now to enable them to return home, and then collect the balance at the ultimate reckoning. The widows, parents, and children who had remained in the old country were judged by the Committee to be just as eligible for relief funds as those who lived in the United States. This action by the MRC differed from Rockefeller's, who thought that no relief whatever was due to the foreign families abroad; and from the coal company itself, whose final settlement with the survivors had been on a rather selective basis.

News of an impending settlement caused an increase in the Committee's paper work. Rose Gaitano of Chicago wrote on behalf of Maria Abbate, a mining widow, who had opted to return to Italy: "I let you know that Maria...has gone to Italy and she brought that check that you have sent to her and she went to one of the banks [in Italy] and they would not pay it for her. They said they could not find Fairmont on the map...." Though the MRC's Status Form had listed Maria as "not clean or thrifty," it still made every effort to pay her $525.00. Some of the Committee's regular work had to be suspended until proper steps, in the form of maps and explanations, had been taken to show the geographical location of Fairmont and Monongah.

Nikulas and Anna Bustin of Hungary asked the Committee to investigate Barbara Tomko. Alex Bustin, their brother, had boarded with the Tomko's until his death in the explosion. He had allegedly given $100.00 to Barbara for safekeeping. The Committee could only reply to this, and similar enquiries, in the negative. If they tried to comply with every such request and trace down the desired information, they would have no time for other matters.

One item, however, the Committee did investigate. Elizabeth Pasquale, sister of Tony who was killed, said she was solely supported by her brother, and thus eligible for relief. It was discovered, though, that Tony had started to work in Monongah only in October, 1907, but that Elizabeth had been there for about a year, keeping house for eight or nine boarding miners. Elizabeth was subsequently stricken from the relief rolls.

Several parish priests wrote from Italy, praising the charitable impulses of the MRC. One problem, however, was that the money already received and anticipated from the Committee produced an economic imbalance in several little villages, creating jealousy among the households. Nicolo Fezioli sent word from Campobosso that Giovanni Colorusso's

widow was in dire straits. Giovanni Fabiani corresponded
from Civita d' Antino: "You can't very well have an exact idea
of the grief brought to the poor family [of Antonio DiMarco].
The aged and invalid parents have lost the only support of a
large family. The deceased [Guissepe and Umberto—both
victims of the explosion] have left also a crippled brother and
two little sisters age 10 and 12 years. All the people simatize
[sic] with the poor family, and the aged mother is almost out of
her sense with grief."

Canon Vincenzo Discegli, Director of the Cathedral of
Detanto, wrote about the widow of Raphaele Cuoccio: "She is
calling for her husband. . . . She and two children are suffering
from want. She was unable to buy mourning clothes, wearing
only a black shawl." Vincenzo Totino, Vice Rector of Santa
Caterina's Church, wrote dozens of letters about the character
of the people who were lost in the mines. He stated that in many
instances the victims' dependents had become public
charges—thus obligating the Committee to help them.

Still another letter from Italy was from the wife of a victim,
asking for relief. Upon investigation, the Committee found
that the man in question already had a wife in Monongah. The
West Virginia wife knew nothing about her counterpart in
Italy. There is no record to indicate which of the victim's
widows received relief.

The Committee continued to write letters and to appeal for
money. Whenever it corresponded with a black person, it
always put in parenthesis after his name, "colored." Thus, a
letter was sent to William Watkins (colored) of Welch, West
Virginia, telling him that his brother Jesse (colored) was a
victim of the mine explosion, and wanted to know if there were
any persons (colored) who were dependent on Jesse (colored).

One of the Committee's appeals in March, 1908, brought a
slap-on-the-wrist from Fairmont President Watson. Over his
private telegraph line, he cabled R.T. Cunningham of the
Committee: "Have heard of some letters sent out by Relief
Committee. . . . Think it is too late to send out appeals at this
time, and they are creating very unfavorable impressions.
Unless Committee have good reason, please see if they will not
withhold these circulars. . . ." A. Howard Fleming replied—
again on the private line—that "about 100 letters to which you
refer, mailed to multi-millionaires . . . so that now it is too late to
recall them. There will be no further appeals or subscriptions."

This exchange only bolstered the general knowledge that the coal company and the MRC were in collusion with each other. Indeed, it may well be that the existence of the MRC took much of the pressure off the Company in terms of its ultimate settlement with the survivors. Of course, no law required the Company to make any settlement; but public opinion did. Thus, the more money collected by the MRC, the easier it would be for the Company. It still had to fight continuing rumors, however, that it had taken full responsibility for the financial welfare of the victims' dependents. These reports were discounted in several letters from the MRC to the Red Cross, to Charities and the Commons, and to Louise Radford Welles, editor of Chicago-based *New Thought*.

Still another story that greatly perturbed the Committee and, obviously, President Watson, was that the Company intended to build an orphanage for all the children left fatherless by the explosion. This suggestion, which received national attention and won much good-will for the Company, came from none other than Mrs. Watson. She had been quite active in relief work, but this time she overdid it. She was told as circumspectly as possible to make no further suggestions— at least in public.

The question of insurance arose on numerous occasions. Who had it, how much, and should it influence the MRC's final settlement with survivors?

Only one hundred fifty of the victims had insurance, and there was a vast difference in this respect between American and foreign miners. The average coverage for the American victim was only $25.00, because "their occupation is an especially hazardous one, and their...contract stipulated that in the event of a general accident or catastrophe, the companies' liabilities would be greatly limited."

For the foreigners, however, the insurance in many instances was quite high. There was still a large immigration to this country in 1907, and the foreign organizations found it profitable in most cases to insure at least the head of the household. These companies, however, did not reckon on events like Naomi and Monongah.

It might be more accurate to refer to the insurors of most of the workers as fraternal orders rather than as "insurance companies." For example, there were nine American Indians in the Monongah explosion, Charles Honaker, the trapper boy, among them. They were provided for by the Improved Order of Red Men and the Degree of Pocohontas. In each instance, the victim had taken out rather extensive policies on his life.

Other groups that had granted liberal insurance policies included The Ancient Order of United Workmen, The Slavish Society, and the German Beneficial Union. Altogether, the insurance liabilities of these organizations came to over $60,000, causing some of them to claim that they faced bankruptcy as a result. Accordingly, they watched very carefully the coroner's inquest on the Monongah explosion. If the Fairmont Coal Company were proven to be at fault, the fraternal organizations intended to petition the national congress for reimbursement of any funds paid to survivors. As it ultimately turned out, the insurors received no help from any quarter.

By early Spring, 1908, the Monongah Relief Committee was ready to make a final judgment in respect to compensating the dependents of the mine explosion. It had received, it believed, just about all the money it was going to, and now it must be divided. From West Virginia had come $43,160; from Andrew Carnegie's Relief Commission, $35,000; Red Cross, $3,478; the government of Hungary, $1,610; the United Mine Workers, $1,000; The American Colony in Dresden, Germany, $104.00; England, $50.00; France, $50.00; Mexico, $10.00; Cuba, $5.00; Pennsylvania, $23,695; Kentucky, $445.00; Alabama, $55.00; and from Alaska, $2.50. These sums did not include the dozens of children's drives around the country for the sufferers of the mine explosion.

An equitable distribution, the MRC decided, would provide each widow $200.00, and each dependent child under sixteen $155.00. This latter figure was ultimately raised to $174.00. Under this scheme, widows with several children could collect, relatively speaking, a great deal of money. Mrs. G.L. Davis, for example, with eight children, received $1,592; Mrs. Mary Santee, mother of six, got $1,244; while Mrs.Blanche Martin and her five children received $1,070. The distribution to foreign countries was: Italy, $43,154; Hungary, $14,257; Austria, $3,763; Russia, $5,176; and Turkey, $800.00.

If the MRC thought its job was done by ascertaining these divisions, it was sadly mistaken. In fact, its task was, in a very real sense, just beginning.

Each time the Committee received a letter from abroad, the

appropriate foreign consulate was usually asked to translate: the Russian and Austro-Hungarian Vice-Consulate in Pittsburgh; the Italian Consul in Philadelphia; and the Turkish Embassy in Washington. The Committee thought nothing about sending dozens of hand-written pages to these agencies expecting, apparently, over-night services. Showing further its linguistic naivete, the Committee had to be informed that there is no such thing as an "Austrian language." The relationship between the MRC and the various legations became increasingly strained in the efforts to compensate foreign survivors of the coal mine disaster.

For example, the MRC contacted the New York banking firm of Knauth, Nachad, and Kuhne to facilitate the payments to the survivors in Europe. The financial corporation agreed to handle the accounts, charging a commission of one half percent, plus 75¢ for each payment. The company's spokesman told the MRC: "...Considering the clerical work which is connected with the execution of such payments, these conditions are very moderate. While our charge is but minimal, we have to pay our European friends quite a high commission...." The firm would send the MRC check to the bank nearest the residence of the beneficiary, who would have to establish identity. Then the beneficiary would be handed the check, less the fees and the handling expenses.

When each check was forwarded, the Committee enclosed with it a declaration: "This amount, please understand, is a gratuity to which you have no legal right or claim, and you are indebted solely to the general public of the U.S. of A. for the same...." It also sent along an admonition: "The Committee is making this payment direct to you with the feeling that you are competent to take care of the same for benefit of yourself and fatherless children, and earnestly recommends that you deposit this money in a strong bank at interest, and that you use as little from month to month as you can get along with."

The recipients usually understood the economic implications of these communiques (except for having to deduct the fees and delivery charge), but very often they were at a loss in respect to the "instructions." Thus, they frequently wrote to the MRC, asking for clarification. This, of course, caused the Committee more and more to rely upon the foreign missions for translations.

The Royal Consul of Italy showed its pique as early as

June, 1908, when it told the MRC:

> If you wish to correspond directly with the heirs it
> would seem a waste of time to write to us. Although
> we are perfectly willing to attend to the entire matter,
> there does not seem, however, to be any reason why
> the same ground should be gone over twice. We have
> all along been anxious to deal with the matter in the
> first instance, feeling that otherwise through a
> multitude of mistakes in names and addresses at both
> ends, it would eventually have to come to us, and it
> would be more satisfactory to all parties if it were
> under our charge from the inception.... But, of
> course, if you prefer to deal entirely with the heirs in
> Italy, you are at full liberty to do so.

But then, at least as far as the Italian Consulate was
concerned, things began to get out of hand. The citizens of Italy
to whom the MRC wrote, lived in remote villages, and they
were usually unable to read their own language, let alone
"technical English." They were not sure, either, in many cases
whether or not they were being duped. This caused the
Philadelphia Consulate to cable Rome with the
recommendation that the relief recipients accept the money
and conform to MRC instructions. It also wrote another letter
to the MRC, this one much testier than the previous one:

> I can hardly refrain from saying to you that I think if
> you had less of what might be termed suspicion
> against the Italian authorities, and if you had
> inquired about the method by which these amounts in
> their entirety could have been sent directly to the
> beneficiaries, I could have suggested a method which
> would have saved the latter the expenses of delivery
> charges, and you the trouble of corresponding. Had
> you consulted me, I could have advised you that a
> draft in lire Italiano made payable directly to the
> widow or parents could have been sent to me, and I in
> turn could have forwarded it to them through the
> public authorities in Italy [rather than through a
> private bank], who would have delivered it without
> the possibilities of any deductions whatever. In this
> way, neither the Royal Italian government, nor the
> Consulate, nor the bankers would have made any
> deductions from the amounts.... It seems
> incongruous that ignorant persons in Italy are
> expected to understand technical letters in the

> English language, while a public committee [the
> MRC] in this country has refused to accept Italian
> certificates summarized by me in English in my
> official capacity.... I would take the liberty of
> suggesting that hereafter when you desire to make a
> payment, you will be kind enough to send the drafts
> payable to the parties, but addressed to my care. I
> shall take much pleasure in forwarding them to Italy,
> with a receipt drawn up by you to be returned in due
> course.

The Committee did not take this advice of the Italian
Consulate, thus compounding an already sensitive situation.
Adding to everyone's troubles was an earthquake in
Campobosso, the Italian region from which most of the victims
came, further unsettling lives and disrupting lines of
communication.

The MRC had its troubles, too, when it tried to deal with
several subjects of the Ottoman Empire who were heirs to some
of the Monongah victims. For one thing, Turkish politics in
1908 were extremely unsettled. In June of that year, Russian
Tsar Nicholas and King Edward VII of England met at Reval
(Estonia) and agreed to intervene in Constantinople on behalf
of the Christian population in Macedonia and other provinces.
This intervention caused a group of "Young Turks" headed by
Mustafa Kemal, to effect several internal reforms in Turkey,
and to pursue "westernization on their own." The Turkish
revolt checked Western intrusion, and ultimately overthrew
the government of Abdul Hamid II.

It was not until February, 1909, therefore, that any
meaningful help could be given to Turkish survivors. A letter to
the MRC, February 13,1909, from R. Ahmed, Secretary of the
Turkish Embassy, spoke of the "New Rule" in Turkey, and his
country's willingness and eagerness to cooperate with the
Committee. The elevation of the Turkish mission in
Washington to embassy status, helped considerably to
facilitate aid to that country.

The Russians were almost as incensed as the Italians in
playing the role of "obedient servant" to the MRC on one hand,
and the beneficiaries on the other. In August, 1908, the
Consulate-General in New York requested the MRC to send
Russian funds to it, to be distributed directly to the heirs
according to MRC specifications. This suggestion the

Committee rejected outright. Thus, in December, 1908, the Russians complained—as the Italians had earlier—that the victims' survivors were illiterate; yet, the MRC and Knauth, Nachod, and Kuhne apparently expected them to read and understand complicated English. This, in the Russian view, was unreasonable. The Consulate-General also wrote of local banks in Russia sometimes pretending ignorance on how to get the payments to the proper people. It strongly suggested again that, to prevent any more delays and possible frauds, the MRC deal directly with it. Somewhere along the way, the MRC was admonished, there must be some kind of faith that the official representatives of the Russian government were honest, and would not divert the payments for personal use.

The MRC acceded to the Russians—at least once. It tried for several months to send money to Mrs. Piers Galskey, who lived in the Lupna Province of Russian Poland, and all attempts had been futile. The Russian Embassy "miraculously" got through to her.

The work of the Committee continued this way until November of 1908, when the Fairmont *West Virginian* printed a story entitled "Monongah Relief Committee Makes Its Final Report, and Adjourns to Meet no More." The article stated that all the money raised had been distributed throughout the United States and Europe. The *West Virginian's* observations, however, were somewhat premature. Letters kept coming in to the Committee, and out, for the next several months. They were handled by Frank P. Hall, on a leave of absence from Citizens Dollar Savings Bank. He carried out these functions for eighteen months. When he completed his work, there were still approximately $400.00 scattered in the MRC's accounts at various "strong" banks. (There were eleven of these altogether, ranging all the way from Fairmont to Bennettsville, South Carolina). The accounts remained intact until 1925. On St. Patrick's Day that year, an explosion ripped through the mines at Barrackville, West Virginia, necessitating the creation of yet another relief committee. The MRC contributed its last $400.00 to the Barrackville sufferers. Thus finally ended one of the United States' best examples of a private welfare program.

The national publicity given to the Monongah Relief Committee focused public attention on the need for federal involvement in mine safety and welfare programs. Though the Committee's work definitely reflected the charitable impulses of the American people, the belief was increasingly evident that relief from major industrial accidents could be handled efficiently and fairly only by federal agencies. When one considers that from all causes, there had been over 20,000 coal mine deaths in the past two decades, the private sector was indeed strained in the matter of relief. One can see why a citizen could conclude, first, that the company involved should be responsible and, second, that tax dollars could at least contribute to the solution. The American public had drifted since the early 1890s toward a greater involvement by the government in areas of industrial safety, wages, and accident compensation. This drift was aided considerably by the "Social Gospelers." Theodore Roosevelt was the first President to pay serious attention to these reformers. The dreadful events of December, 1907, bolstered the reforming impetus, and helped produce a safety conscious and regulation minded public. The President's renewed plea to Congress to create a Bureau of Mines was just one example of this reformist activity.

One procedure of the MRC that deserves special attention was the decision to mete out contributions on a basis of equality regardless of any other benefits that may have accrued to the survivors, and of paying little attention to the moral and political persuasions of the recipients. "Suffering is suffering, no matter the person, no matter the cause," seemed to be the MRC's attitude, and while it did tend sometimes to be morally righteous—to let the recipient know that "they who give are superior to those who receive"—it tried its best sincerely to give assistance wherever it was needed. In most cases, it succeeded.

Another noteworthy procedure was in dealing with the foreign survivors. It is astounding that an American committee in 1907 would be this solicitous of Italians, Turks, Austrians, Poles, Russians, and others, at a time when unlimited immigration quotas produced so much economic and political hatreds throughout the United States. Labor unions, particularly, opposed unrestricted immigration, for foreigners competed for their jobs. Companies, on the other hand, often viewed the newly arrived immigrants as cheap sources of labor, and exploited them.

The Committee laid aside national rivalries and treated all survivors of the Monongah explosion with equality. This decision ran counter to the ideas of "philanthropists" like John D. Rockefeller, Jr., who believed no consideration at all should be given to the foreign heirs. One can attribute this procedure of the MRC purely to charitable inclinations, for the practice itself was probably disagreeable to the industrial segment of the United States—the very segment that supported the Relief Committee to begin with.

The Committee's use of a New York banking firm to get payments to the European survivors, and its disdain of the official representatives of numerous foreign governments, could indeed have had international repercussions. There had been instances in the past when foreign authorities protested state and private activities that hinted at having anything to do with the foreign policy of the United Sates. The MRC, however, was involved only in charitable work; thus, the disagreement between it and the foreign consulates was one of procedure rather than of substance. Certainly, there was no question here of violating the Logan Act, the Law of 1798 which made it an offence for a private citizen to interfere in a controversy between the United States and a foreign country. The foreign consulates *could* have made it difficult for the MRC to go about its work of aiding European survivors. There is, however, no indication that the matter ever became serious enough for the consulates to contact the U.S. Department of State.

The MRC was only one of several such committees that operated in the early part of the twentieth century. The San Francisco earthquake, for example, had produced several general appeals for help throughout the nation. Following hard on the heels, as it were, of the San Francisco Relief Committee, was the formation of the MRC. All of these operations severely taxed the willingness and the capabilities of the general public to keep "footing the bill" for one disaster after another. That the public was tiring of such matters can easily be seen by contrasting the great publicity enjoyed by the MRC, and the relative obscurity of the Darr Relief Committee, established after the Pennsylvania tragedy on December 19.

On the whole, Americans have always been a generous lot, giving money not merely because they have it to give, but because of a philosophy based in part upon Puritan influences that demands the alleviation of suffering, and the defense of helpless people. The frontier experience, among other things, had taught many lessons in mutual assistance to the American people. This inheritance cropped out more than once during the dreadful month of December, 1907—but it may well be that this was the last time to such a high degree and intensity. If this statement is true, then the dreadful month marked an important transition in the cultural and social outlook of the American people.

Sources

Birmingham *News*, December 7, 1907.

Byington, Margaret F. "The Monongah Relief," *Bulletin,* The American National Red Cross (Vol. III, No. 2), Washington, D.C.

_____. "The Relief Problem at Monongah," *Charities and The Commons,* January, 1908, p. 1451.

Elliott, Mabel A. *The History of the American National Red Cross.* Volume XX-A, Washington, D.C.: The American National Red Cross, 1950

Fairmont *Times,* December 17, 1907.

Fairmont *West Virginian,* November 23, 1908.

Haas, Frank. *The Explosion at Monongah Mines.* Bulletin No. 11, Fairmont Coal Company, 1910.

History of the Monongah Mines Relief Fund. Fairmont, West Virginia, 1910.

MSS. American National Red Cross, Washington, D.C.

MSS. West Virginia University; Morgantown, West Virginia.

Pittsburgh *Dispatch,* December 7, 8, 1907.

Perspectives-I

The Teens and Twenties

Did things improve after the dreadful month of 1907? For an answer, consider these points:

In October of 1913, on the 22nd day, an explosion ripped through Stag Canon Coal Mine in Dawson, New Mexico; 263 of its 300 employees were killed. This was a "model mine" according to its owners, Phelps-Dodge Company, of New York City.

"...It was supplied with all the latest appliances to make it safe," averred one company official, who cited the occasional use of electrical detonators for shots, fired from the outside. Dawson's mayor, T.L. Kinney, tried to soothe the gathering crowds by claiming that the mine was constructed in such a way to permit "the least possible danger from such explosions."

Was one of the "latest appliances" of which the company official boasted the dust spray that was not effective beyond six feet? This in a mine that was well known for its dryness and dustiness?

Did the company talk about "safety practices" only to the public, and at the same time allow open lights in the mine, and also permit shots with any type of explosive to be used? So it would appear.

The explosion, which occurred at 3 p.m., raced along haulage roads and sped into nooks and crevices, dealing out instant death everywhere it touched. Mayor Kinney's claim of "superior construction" to minimize explosions was about as truthful as the company's plea of using "latest appliances."

A disagreement arose between the miners and Rees A. Beddow, New Mexico's chief mine inspector. Beddow reported, and all the company officials agreed with him, that an overcharged shot in Room Twenty-seven blew the coal outwards for forty feet, creating much wind, and stirring up and igniting the dust. Gas, he said, played little or no part in the catastrophe. He maintained that his inspection of just one week before only found minute traces of gas.

Not so, retorted several workers. A large pocket of gas in the coal was tapped, and was set off by an open torch. There had been some reports of gas in the recent past—overlooked or ignored by the officials. Whatever—dust, gas, or both—it appears that the explosion occurred from a combination of company negligence and worker carelessness. The lessons that could have been learned from past experiences apparently came to nought.

Naturally, the crowds grew quickly, hampering rescue operations, made up of miners from the neighboring fields. The same morbid scenes of Naomi and Monongah were replayed as friends and relatives frenziedly tried to get to their loved ones. Ironically, they were pacified—at least partially—when they learned that a man named Frank McDermott was one of the imprisoned miners. He was the mine superintendent, and surely all would turn out well with such a high ranking official inside. The crowds, said a reporter, had "confidence, as have mine officials, in the superintendent's resources in such a plight."

Frank McDermott perished on that miserable day of 1913.

And so it went. Amid the chants of Austrian widows, the cries of Mexican women, the groaning of Greeks, and the sobbing of Americans, the funerals were conducted. Lives were re-fashioned. The mine was re-opened. New men took the place of those who had died....

In February of 1923, on the eighth day, an explosion ripped through Stag Canon Coal Mine in Dawson, New Mexico; 120 men were killed. Officially, that is; records indicated that 140 had checked into work that morning. This was a "model mine" according to its owners, Phelps-Dodge Company, of New York City.

The explosion occurred at 2:30 in the afternoon. Two cars loaded with coal were de-railed, and were dragged for one hundred feet before their locomotive stopped. Dozens of support timbers were displaced, causing huge clouds of dust. Simultaneously, the mine's wiring was severed by the wreck, producing electrical arcs which detonated the dust. The explosion raced along haulage roads and sped into nooks and crevices, dealing out instant death everywhere it touched.

W.D. Brennan, manager of the mining company, said

View of portal, Stag Canon Coal Mine; Dawson, N. Mex. February 8, 1923.

Miners standing around concrete slab blown from Stag Canon Coal Mine; Dawson, N. Mex. February 8, 1923.

there had been an inspection the day before the explosion. The dust was down, he said, and the mine was in "excellent condition."

How, then, does one explain that for five days before the explosion, there had been no sprinkling of dust whatever because all the water pipes were frozen? If anything, there was more dust now than ten years before, when the mine had blown up. This was also February—the 1913 explosion was in October—and cold weather, as we have previously noted, helped materially to increase the danger of dust explosions. Dust, therefore, was thick in the haul ways, in the rooms, in the crevices, and even on the cross-bars on the roof. Yet, some officials kept saying that the tragedy was a "mystery."

These 383 men, joined by five others from an explosion in Dawson's Stag Canon Mine on April 14, 1920, were among the legions who, the officials kept saying, worked in "safe mines," with the "latest" in safety equipment. The records show, however, that the officials were much more adept at making profits than in saving lives.

Pennsylvania was no stranger to the horrors of coal mine explosions. On November 6, 1922, the Reilly Mine at Spangler demanded the lives of 79 of the 94 miners in it.

"Bodies were scattered through the workings for more than 500 feet," reported one survivor.

"The sad part of it," said a rescuer, "is that at least 20 of the victims were within 100 feet of fresh air when they were overcome by gas."

The rescuers came to a brattice made of mine cars. A legend was burned on one of the cars by a miner's lamp: "There are 20 men behind this." There was no hope for them, though. The air was so bad at that point that the rescuers' pet canary, "Sally," died. Canaries had long been used in England and Wales to help miners spot the hazardous places. If the air was impure, the canary died very quickly.

A lot of people spent that day at Spangler looking for Pat Flanagan, the fire boss. He had gone into the pits at three that morning, and had declared them safe. This, despite a previous record of note:

"The mine had been rated gaseous in 1918, but at the insistence of the new operators (Reilly) it was rated as non-gaseous although a fire-boss (Flanagan) was employed and men were burned by gas on at least 4 occasions."

A story circulated through the waiting crowds that Flanagan "was known to have exploded small accumulations of gas" in the past in order to clear the mine for working, a practice not openly permitted by the company. Whether or not this "charge" against Flanagan was true, was never really proved.

The explosion was caused by a vast accumulation of gas coming into contact with an open lamp. Many miners not killed outright by the explosion were felled by the deadly afterdamp. A survivor testified:

"From 7:30 until about 2 O'clock, we stayed in the chamber. Older men were getting faint. Young fellows were holding out pretty well. Some of the boys were praying.

"Finally the afterdamp worked its way around and under the door. There was only one thing to do. I told them we would have to run for it. We opened the door and ran down the heading. It was a tough dash, stumbling over bodies. We began to choke. Some of the boys fell, but those of us who had any fight left kept on.

"Only the young men made it; I'm afraid all the older men lost their lives."

If this explosion at Spangler had been on any day but Monday, the fatalities would have been higher. The miners never really recovered from their week-ends until Tuesday mornings, when 200 of them reported for work.

Incidentally, Joe Flanagan, the fire-boss, was never seen again. Apparently, he re-entered the mine later that morning, only to become its victim.

Mather, Pennsylvania is almost across the state from Spangler. The rich coal fields, however, assured that the little community would have no immunities from tragedy.

The afternoon shifts were changing at Mather No. 1 on May 19, 1928, when the explosion of gas and dust occurred. Of the 209 men in the mine, only 14 emerged safely. This was a "model mine," said by the officials to be in "excellent condition."

One of the locomotives inside the mine, so it was reported, used a storage battery—one that after sitting for some time frequently shot off sparks upon ignition. The coal company, Pickands-Mather of Cleveland, outlawed the use of these batteries, but obviously had not enforced its own regulations. Again, it appears that safety rules were announced for the benefit of mine inspectors and the public, while inside the mine itself, it was "business as usual."

Alabama's position as the most industrial state in the South is enhanced by the thick veins of coal running through its earth. On November 23, 1922, at the Dolomite Mine near Birmingham, catastrophe struck, killing 90 men.

The event was, to some extent, like the one at Monongah. Four "trip" cars broke loose at the tipple, and raced uncontrollably down Number Three Slope. It was on a thirty degree pitch, and it was 800 feet long. The runaways produced huge amounts of dust. They also sharded the 3,300 volt cable running down the slope. The arcs from the cable ignited the dust. Some survivors spoke of "streaks of flame" like lightning bolts shooting out of the mine's mouth. So severe were these that the tipple itself was engulfed and consumed; many casualties in this calamity were people working on the outside of the mine.

This mine, only recently inspected by state and federal personnel, was declared "safe." Therefore, the experts claimed that the explosion was a "freak."

Perhaps so, especially since it was originated by runaway cars. But why did the runaways *have to* activate hundreds of tons of deadly coal dust? Proper sprinkling methods and ventilation systems would have gone far to help the officials at Woodward Iron Company, owner of Dolomite, retain their much flaunted "safety" record. Also, which were they interested in more; safety or profits? The first thing the miners saw each day as they descended into the pit was a huge sign, put up by the company, that read:

"Last month we broke our record for tonnage. Let's make it 52,000 for November."

An additional mar in this "safety" record was the mine's telephone being located two miles away from the pit, in the company office. Thus, it took an unconscionably long time before word reached the outside that an explosion had occurred. It was the happy fact that the ventilation system was not blown out that saved so many lives (an estimated 477 men were in the Dolomite Mines at the time); not the speed with which outside help arrived.

The officials had always claimed that Dolomite was "gas free." This may have been true—at least to some extent—under normal working conditions. But, read what M.D. Wilson had to say about the gas, just after the explosion:

"The gas began to get worse and we tried to shut it off first by closing a door, but it was no use. The gas got terrible, and all about me men were praying. I prayed a little myself, to die, I tried to choke myself to death, but could not. If there had been a way to end my life, I would have done it, for my suffering was so great. Finally, I lost consciousness."

Mr. Wilson was rescued.

Several of the Dolomite mines were connected to each other by deep underground passages. While these connections did not enlarge the tragedy on this particular occasion, the lesson of Monongah should have shown how careless, negligent, and stupid it was of a company to allow such things. The connections certainly did negate any Dolomite claims to "safety."

Another Alabama mine, always known for being "highly developed with every modern and safety appliance," blew up two weeks before Christmas in 1925. This was the Overton Mine just out of Birmingham and it put 53 men into their graves.

Immediately upon hearing of the explosion, the officials stated that the mine had been "properly safeguarded, that safety appliances had been provided, that the operations were in the hands of men well versed in mining." Yet, that very morning, the fire-boss had found large quantities of gas in four different places, and this gas was supposed to have been cleared before anyone was permitted inside the mine. About 10 that morning, two miners were sent into Room Seven to see if the roof needed additional timber supports. They carried carbide lamps with them, and a goodly supply of matches. They undoubtedly had some trouble seeing where they were, and turned up their lamps. The resultant explosion picked up the large quantity of coal dust, and delivered it to every part of the mine.

One survivor of the disaster was a black man, blinded by the blast. He retained his composure, however, and grabbed fast onto the tail of a mule which was, of course, heading for the exits. The mule dragged him to safety. He said, after the ordeal: "I knew I'd get out with the mule, if there was any gettin' at all."

The Overton Mine was owned by the Alabama Fuel and Iron Company, with Charles DeBardeleben as President. He went into the Overton pit himself, trying to rescue his workers; all of his other nearby mines, however, kept fully operating, and he complimented the men in them because "the coal production of the company has been kept normal." The explosion occurred on December 10, a Thursday. The last body, that of John Rice, was recovered on the 16th, a Wednesday. Full operations in Overton Mine were resumed the next day, Thursday. Except for the shattered lives, everything returned to normal; and everyone could wait now for the next explosion in somebody else's "model" mine.

Historically, West Virginia is the most notorious state in the Union for killing its coal miners. This is true for a variety of reasons. For one, the owners steadfastly fought governmental regulation and attempts at unionism. Also, the workers were subjected to wages directly tied to production. Very frequently, this system caused miners—especially if they needed some extra cash for a special purpose—to resort to dangerous methods. On into the teens and twenties, it was as though Monongah had never happened. The surge for the almighty dollar by the companies and, to a lesser extent by the miners themselves, provided the Grim Reaper with an insurance policy against unemployment.

In the Spring of 1924, on April 28, the mine at Benwood, West Virginia blew up, taking with it 119 men. There were no survivors. This was a "well run" and "safe" mine, with "no" problems relating to gas. Yet it was gas that caused the explosion on that fearful morning.

Two fire-bosses had reported the mine free of gas. A miner, however, so it was surmised, found an extensive fall, one that measured eight feet by twenty-two. Apparently, thinking that chief fire-boss J.T. Poyle had declared the fall to be gas-free, the miner examined it, and lit a match to get a better view. The mine was dry and dusty, so the resultant explosion traveled very quickly to all of its other parts.

An outside witness reported that white smoke burst from the entry, followed by huge sheets of flame. The mine's mouth then began to vomit angry clouds of blackdamp, keeping rescuers at bay for several hours.

Finally, when conditions did permit rescuers into the mine, large crowds gathered on the mine grounds. Included in them were many foreigners, unable to understand English. The women, so it was reported, "stormed and wept...and tried to understand what they suspected to be a plot to keep them from aiding their loved ones." Adding to the complication was the company report that 110 men had checked into the mine that morning, and the general knowledge of the crowd that there were "several other men in the mine, who entered without officially checking." (As we saw with Monongah, this was entirely likely. Young sons and nephews were frequently taken into the mines by the regular workers to help dislodge coal in narrow places. And with wages tied to production, there was always the tendency for miners to get as much individual help as they could. Therefore, it can easily be surmised that in *all* these accidents, the *real* fatality rate was higher than the one put out by officials).

West Virginia Governor, Ephraim Morgan, offering his "personal aid" in the rescue efforts (though he did not define the phrase), promised a "full investigation" into the causes of the explosion. This statement was followed by a declaration from L.M. Scott, President of Wheeling Steel, owner of the mine, that an "official enquiry" would be undertaken. This combination of platitudes from both politicians and company officials assured, as it always had in the past, that the matter would drop right there, and that profits would return to normal just as quickly as possible. And so it went.

The next major West Virginia tragedy was in Everettesville, near Morgantown where, on April 30, 1927—almost three years to the date from the holocaust at Benwood—an explosion ripped through Federal Mine Number Three, killing 97 men. This was a gas and dust explosion which occurred at 3:30 p.m., just as the day shift was preparing to quit work. What caused the explosion in this "well regulated" mine? The best estimate is that an ostensibly out-lawed open-type storage battery locomotive "ignited an accumulation of gas." Mine companies were forever saying that they did not permit certain types of equipment in their mines; coroner's juries were forever listing these "non-permissible" tools as leading causes for explosions.

The Everettesville mine, owned by the New England Fuel and Transportation Company, was fiercely non-union. That is one reason why the company always hailed it as a "model mine," one that used "only the latest equipment." It wanted to show that non-union mines were better places to work in than those dominated by the UMW.

General view of mine wreckage; Everettesville, W. Va. April 30, 1927.

Breakage of mine tipple; Everettesville, W.Va. April 30, 1927.

Perhaps its non-union status also explained why the majority of workers in the Everettesville mine were black; and a substantial portion of these came from Alabama. They had moved North looking for higher wages.

Flames shot from the mine's mouth, killing two men on the tipple, and badly burning several others. Continued steam and smoke from the mouth indicated to rescuers that fires raged within. It was estimated that the miners were back at least 3,400 feet.

The explosion occurred Saturday afternoon; twenty-four hours later, West Virginia's chief mine inspector, R.M. Lambie, called off all rescue efforts. He said, "It is a foregone conclusion that all the...men in there are dead. They couldn't possibly be living in that gas."

When he withdrew the rescue workers, Lambie ordered guards to keep everyone at least 200 yards away from the mine. He believed, he said, that fresh fires and new explosions were imminent.

Lambie was criticized in some circles for stopping the rescue operations. Whatever his reasons, this was the shortest time ever given to trying to recover survivors and bodies. The evidence indicates that Lambie acted out of concern for rescuers and a belief that there were no survivors. His actions had nothing to do with the non-union status of the mine, and the fact that most of its workers were black.

A public funeral was held for the unrecovered victims. Simultaneously in Fairmont and Morgantown, words were said over the body of a victim, who symbolized all those who were lost, "with much the same significance as military honors for unknown soldiers."

The Red Cross and Salvation Army remained on the scene long after the explosion occurred, caring for the widows and orphans and trying to comfort them. And so it went.

Other states besides the four that have been discussed were also visited by horrible coal mine explosions in the twenties. Those with fifty or more fatalities were:

. Kemmerer, Wyoming; August 14, 1923. A blown-out shot, said the company, caused this explosion which killed 93 men. Later, however, it was learned beyond doubt that the company's fire-boss tried to relight his lamp with a match in an area filled with gas. The explosion, propagated by dust, spread to every part of the mine. Those not killed immediately were

General view of a West Virginia coal mine.

soon destroyed by blackdamp. Nevertheless, there were 36 survivors; they survived primarily because they remained at their work stations until rescuers arrived, rather than in a panic trying to run through the blackdamp to safety.

. Castle Gate, Utah; March 8, 1924. The fire-bosses did not report all the gas in this mine, especially that which accumulated in roof "potholes." These were holes left after shooting coal from the top of the shaft. As in the Kemmerer explosion, a fire-boss tried to relight his lamp at one of these gaseous "potholes." It killed 172 men.

. Sullivan, Indiana; February 20, 1925. Again this explosion—at 10:45 a.m.—occurred from gas in a mine that had been declared free of it three hours before. Fifty-two men were lost.

. Farmville, North Carolina; May 27, 1925. According to early reports, 59 men checked into the mine that morning, although 71 miners' lamps were out. Later, the official report listed 53 fatalities.

. Wilburton, Oklahoma; January 13, 1926, 91 killed. Three who weren't owed their lives to superstition. They had refused to enter the pits on the 13th of the month. Gas leaking from the roof was cited as the cause of the Wilburton disaster. Only a few hours before, two fire-bosses had proclaimed the mine to be gas-free.

. McAlester, Oklahoma; December 17, 1929, 61 killed. An open light, either from a match or an arc from the numerous fan and locomotive motors in the mine caused the explosion. As in most other cases, it was spread throughout the mine by thick layers of coal dust.

The fatalities in the major explosions of the 1920s came to 1,281. The "minor" explosions, those that did not receive so much attention, produced 1,088 victims for the decade's total of 2,369. The dreadful month of December, 1907, had obviously not made much impact.

Did things change *now,* after the carnage of the twenties? Yes, they did—to some extent—though major and "minor" explosions continued to take a heavy toll. Did the change occur because everyone was sickened by so much loss of life? Partially, yes. The greatest reasons for change, however, were, as we shall see, something else altogether.

Sources

Atlanta Constitution. October 23, 1913; November 7-10; 23-24, 1922; February
   9-10, 1923; August 15, 1923; March 9, 1924; April 29, 1924; February 21, 1925;
   May 28, 1925; December 11, 1925; January 14, 1926; May 2, 1927; May 23,
   1928; December 18, 1929.
Birmingham *News.* November 22-24, 1922; December 10, 12, 14-16, 1925.
Birmingham *Post*, November 23, 1922; December 11-12, 1925.
Humphrey, H.B. *Historical Summary of Coal-Mine Explosions in the United
   States, 1810-1958.* Washington: Bureau of Mines Bulletin 586, 1960.
Raton (New Mexico) *Range.* November 4, 7, 11, 1913; October 24, 28, 31, 1913;
   February 9, 13, 23, 27, 1923.
The Daily Tribune (Johnstown, Pa.). November 6, 7, 8, 9, 1922.
The New Dominion (Morgantown, West Virginia). May 2, 1927.
The Waynesburg (Pa.) *Republican.* May 24, 31; June 7, 1928.
The Wheeling (West Virginia) *Intelligencer.* April 29, 1924.
Tuscaloosa (Alabama) *News and Times-Gazette.* December 11, 1925.

Chapter Four

The Spreading Specter:
1907

In the hectic, frenzied scramble for industrial properties at
the turn of the twentieth century, no state was more significant
than Alabama. Its industrial potential had long been
recognized; and, in fact, its economic capabilities had caused
the Reconstruction Congresses of the 1860s and 70s to treat
Alabama with considerably more deference than the other
Southern states.

Alabama's industrial land boom produced vigorous, cut-
throat competition among numerous strong-willed
individuals. It was a system of buying and leasing, and of
taking options on as much coal and coke lands as possible.
Frequently, the boom reached the fabulous proportions of New
England and Mid-west land speculations of mid nineteenth
century. The person not willing to take the calculated risk of
industrial speculation soon dropped out altogether. But for
those who had staying power, the monetary rewards were
almost always tremendous.

One of the survivors was an enigmatic Alabaman with the
unlikely name of Graton B. Crowe, whose father had once been
the Governor of New Mexico territory. Crowe was born in
Marion, Alabama, and early in life he demonstrated an
aptitude for medicine, studying that subject at an Alabama
school named Old Briarfield.

Crowe also had a propensity for poker, and other card
games. As was the practice of that day, he frequently went off
into the woods with his friends and acquaintances for an
afternoon of gaming. On one such occasion, an argument
developed, and Crowe shot and killed Benjamin F. Glass. He
was tried for murder, but acquitted on the grounds of self-
defense.

Wanting to further his knowledge of medicine, Crowe
studied for a few years at the University of Edinburgh, in

Scotland, becoming when he returned to Alabama, a country doctor in Bibb County. By now, however, politics fascinated him, and in 1896, he was a gubernatorial candidate on the Populist ticket. Being thoroughly defeated in the contest, Dr. Crowe decided once again to leave the United States. He traveled around Europe for a year or two, taking several special courses in geology from various European universities. This geological training put Crowe in an excellent position to speculate on industrial properties. By 1902 he was the president of half a dozen coal mining companies in Alabama, and was recognized as the fourth largest operator in the State. Along with the Watsons in West Virginia, he had become one of America's wealthiest people.

During his travels abroad, Dr. Crowe grew fond of European culture, especially its music. One story that fascinated him was *King Rene's Daughter,* written in 1845 by Danish author Henrik Hertz. King Rene had a beautiful daughter named Yolande. The play was frequently performed before European audiences, and later, Peter Ilich Tchaikovsky based his opera *Yolande*, on the subject. Crowe knew also that "yolande" was Latin for "violet."

It is ironic that Dr. Crowe used this literary, musical and linguistic motif as the name for a little mining village in the northeastern part of Tuscaloosa County, Alabama, thirty-five miles south of Birmingham. Since he steadfastly refused to hire any union workers, and since the place was named so beautifully, Dr. Crowe tried to turn Yolande into a model mining community. The white, green-trimmed houses, where the miners lived with their families were on top of a hill, giving an air of picturesqueness and healthful living—unusual for mining camps. The wages at Yolande were above average, and it used all the latest mining equipment. Thus, Yolande—like Monongah—was regarded as one of the best and safest mines in the country. Yet it exploded on the 16th with terrible loss of life, becoming the third calamity in the dreadful month of December, 1907.

Superintendent T.C. Huckaby inspected Yolande on that Monday morning. He was well pleased with the condition of the rooms, and complimented several miners for their attention to safety factors. Shortly after 10, he came out of the pit. Whistling a tune, Huckaby saddled his horse, happily looking foward to a day of hunting. As he rode off, a low rumbling noise caught his attention, followed immediately by a loud explosion. Huckaby looked up and saw a horrendous cloud of dust, and also timber fragments flying through the air, some landing as far away as the depot, across a ravine, three hundred feet away. Simultaneously, several small buildings collapsed. Huckaby realized at once that he would still go hunting on this day—but not for deer.

For many residents of Yolande, the great noise put them in mind of Virginia City, a near-by mine where over a hundred laborers had been killed two years before. Would it be that bad this time? How many miners had gone into Yolande pit that morning? Today was Monday, so maybe not as many as usual. But enough. Yes, indeed, enough. The crowds gathered around the pit, and at once grew almost to unmanageable proportions.

Rescuers came, but were held back from entering the mine by the intense heat and the blackdamp oozing from its mouth. The local rescuers were soon joined by miners from Davis, Brookwood, Connellsville, Blue Creek, and other mines in the district. By noon, fourteen badly burned survivors had crawled to the surface. Their descriptions of the inside left little hope that the other miners—some 1500 feet down—could be alive; information that brought fresh wails and moans from those in the crowd who had men below.

At his office in the First National Bank Building in Birmingham, Dr. Crowe received word of the Yolande tragedy. He immediately did two things:

. Ordered a special train to take him and other company officials to the scene.

. Issued a statement that early reports of casualties were exaggerated. More like fifteen or twenty, said he; certainly under forty and definitely not in the seventies and eighties, as some journalists had suggested. This seemed to be a common practice in every mine accident. The first thing the company president always did was downplay the number of fatalities. He did this whether or not he had actually been on the scene.

By mid-afternoon, hastily constructed ventilation systems pushed cool, fresh air into the Yolande pits, but it was used only by the rescuers; there were no more survivors. As the day wore on, additional bodies were brought to the surface, causing a fresh outcry from the crowd at each arrival. There were few foreigners here, compared to Naomi and Monongah; and the general understanding of English made orders easier to follow, facilitating the rescue work. The victims seemed mostly to be young, single men in their early twenties; thus, no huge population of orphans was created by the disaster. There were enough mothers and fathers and other family members, though, deeply to lament the loss of loved ones. As night began to fall, a cold, slow drizzle set in which dispersed many of the curious on-lookers. For those with men in the mine, however, the rain was hardly felt.

Despite President Crowe's staunch denials, rumors persisted that over one hundred men had checked into the mine that morning. Although he officially stated that casualties would be between twenty and thirty, at 2 that afternoon, the Gaudin Undertaking Company of Bessemer had forty caskets on the way to Yolande. It announced that its representatives would stay on the Yolande grounds until every victim had been accounted for. The funeral wagons added to the gloominess of the day, making the survivors realize that Crowe's computations were not correct.

By Wednesday, two days after the explosion, all the bodies had been recovered—officially fifty-seven of them, and thus began once more the doleful job of a coroner's jury. Why did J.J. Morrison, J. Smedley, and fifty-five of their pals perish on that fateful morning of December 16, 1907? A blown out shot, said one official, caused by a miner's carelessness. Excessive gas, reported but not dealt with, averred a seasoned miner. The cause of the explosion, along with its exact number of victims would be heatedly discussed for months to come. The only *true* reality was that of widows and children, roaming about, wondering what to do.

Violent death is always attended by a certain amount of irony. Nowhere was this more clear than in the case of Matthew and Andrew Stoves, father and son. They came from Irwin, Pennsylvania, in Westmoreland County. A year ago, they heard of the above-average wages at Yolande, and of the safety-designed equipment in use there. Perhaps Matthew could support his wife and five minor children if he and Andrew traveled down to Alabama. Their year in the South seemed to confirm their optimistic outlook, especially when northern mines like Naomi and Monongah began to blow up.

But then, the Stoves did not reckon on the events at Yolande, and they were lost on that doleful day. Dr. Crowe paid for their transportation back to Pennsylvania, accompanied by a company representative. The entourage reached Irwin on the 22nd; the funeral was held the next day. Only the widow and the weeping children attended. No neighbors were there, or friends; no community attention whatever. The reason for this absence was not indifference; far from it. The reason was that the Stoves were correct about the frequency of explosions in the northern mines. Westmoreland County had its own suffering to do, for it was reeling from the worst mine disaster in the State's history. On the 19th, the Darr Mine at Jacob's Creek blew up, contributing another 239 victims to the steadily rising statistics of that miserable, final month of 1907.

Westmoreland County, Pennsylvania, contained some of the richest coal and coke lands in the world. From the mid 1850s to the turn of the century, the area was a veritable beehive of activity as one fortune after another was made from "King Coal." The wealthiest and most powerful enterprise in Westmoreland County was the Pittsburgh Coal Company. Like United at Fayette City, the Watsons at Monongah, and Crowe at Yolande, it used only the most modern equipment in its "up-to-date" mines.

Scene at Darr Mine; Jacob's Creek, Pa. December 19, 1907.

The pride and joy of the Pittsburgh Coal Company was the Darr Mine, one of five that it operated as part of the old Connellsville Coke Fields. Opened in 1899, Darr was located thirteen miles northwest of Connellsville and forty miles southeast of Pittsburgh. It was on the west side of the Youghiogheny River, directly opposite the little village of Jacob's Creek, and adjacent to another tiny hamlet, Van Meter. On the Darr side of the river ran the Pittsburgh and Lake Erie Railroad; on the other side was the main line of the Baltimore and Ohio.

Most of the Darr miners lived across the river, at Jacob's Creek. They got to work each day by a "sky ferry," a woven metal basket attached to a steel cable. The river's current was too swift here for boats to operate, and the only bridge was at Smithton, two miles to the North. The basket held only six people, so it was used extensively everyday. To get across the Youghiogheny, the miners allowed the basket, or car, to run out halfway over the river; then by winding a wire on a windlass, they drew themselves to the bank. Once landed, the miners still on the Jacob's Creek side pulled the car back to them, and repeated the process. In this way, dozens of men could be transported to Darr in only a short time. On the Darr side of the river, the miners did not have very far to walk to get to the mine's mouth. They only had to cross the railroad, and walk a few feet to their destination. The narrowness of the Darr grounds played a significant role in the disaster that befell its citizens on the 19th.

The Darr slope was approximately 2½ degrees, going back into the mountains some 5,000 feet. Hauling was done mostly by animals though electricity had been installed in some of the mine's areas. Chain type coal cutting machines were employed here, each incision of which was six feet deep and four inches high. A giant Vulcan fan, six feet wide and twenty-four feet in diameter, pushed 110,000 cubic feet of air into the mine every minute. Despite all of this ventilation, it had long been known that Darr was a gaseous mine.

Darr employed about 450 men, most of them making up the day shift, with only minimal operations at night. This number was roughly equal between native American miners, and foreigners who came primarily from eastern and southern Europe. These latter were at Darr essentially for the same reasons for the foreign presence at Monongah. They had come, generally, as a part of the immigrations of the time; and specifically, as strike breakers brought in by the operators at the opening of the twentieth century.

As at Monongah, the Darr foreigners were keen to observe the religious rites of their predominantly Eastern Orthodox Church. Hence, on December 17 and 18, the Darr mine was closed in observance of the celebration of St. Nicholas. The festival this year was so popular and enjoyable that when Thursday the 19th arrived, only a few Europeans were ready to return to the pits. Many of them were still drunk from all the revelry; others had fearful hangovers, causing them to prefer losing a day's pay than to endure these cranial tortures.

The newspapers and other reports of the day called the rituals of St. Nicholas by various names. According to some, it was a Russian holiday; to others, it was a Hungarian festival; to still others, the day belonged to the Greeks and Poles. It was, of course, not limited to any one country; rather, it was a most joyful Eastern Orthodox observation, attracting the attention of numerous nationalities. Well that it might, for it saved the lives of two hundred men: the majority of the victims of the Darr explosion were Americans who were waiting until the 25th to celebrate their own good St. Nicholas. This irony of the different holidays, as we shall see, was to have some unpleasant repercussions.

One miner at Darr was sixty-year-old John Campbell, an American, who left the Eclipse Mine at Finlayville, Pennsylvania, a few months before because it was unsafe. Each day punctually at 11:30 he came out of Darr, and ate the noon-day meal with his wife at their little home in Van Meter. Mrs. Campbell was a coal miner's wife through and through, for her previous husband was John Lester who died a few years ago in the near-by McDonald Mine.

At 11:30 she realized that John was late for dinner. She headed toward the front door to see if he was on the way when the house began to shake. At about the same time, a great rumble, like a sudden clap of thunder, caused her to know at once what was taking place. A reporter for the *Greensburg Press* described her reactions: "... She went to the door and saw great columns of smoke belch from the mine.... She had been a miner's wife too long not to realize the importance of that rumbling. She threw up her hands and gave one despairing scream and ran with a few of the neighbors to the mouth of the mine. There in agony she watched and waited, wringing her hands and praying, hoping against hope until someone gently led her aside and told her that her husband's body was the first one found." She later told reporters that she and her husband had talked frequently about the gaseous condition of the Darr mine. Campbell had gotten almost to the mine's mouth, when the deadly afterdamp overtook him. He was found in a sitting position, holding his head in his hands, trying to evade the poison.

Frank Ballantine managed the Darr commissary, a few hundred feet from the adit. He and his customers of the moment were hurled against walls that began to crumble, resulting in numerous injuries. John Landway, atop the Darr tipple, was blown from it, miraculously escaping with only superficial injuries. That was not the end of the tragedy, though, for Landway: his father was in the mine. Across the river at Jacob's Creek, the ground shook and buildings swayed. Seasoned residents of the mining camp looked up with knowing eyes; many celebrants of St. Nicholas became instantly sober.

The crowds gathered; they came from Jacob's Creek, Van Meter, Banning, Connellsville, Wickhaven, Smithton, and some even from Pittsburgh. The morbidly curious mingled with the mourners, setting up a din of strangely mixed sounds and emotions. Crowd control was easier here than at Monongah, for the narrow strip of land between the Youghiogheny and the mine could easily be roped off. Nevertheless, several police reserves were on hand, and talk of mobilizing the National Guard was heard with increasing frequency. Another reason for a lack of multitudinous congregations was the huge amounts of blackdamp—more even than at Monongah—that continued to pour from the mine. Some officials openly predicted that it would take two days before rescuers could safely enter the Darr pits.

Naturally, there was much speculation in the crowds about the cause of the disaster. This was the fourth major coal mine explosion during the month, and rumors were plentiful. One explanation had it that Darr was still connected to Port Royal, a mine that had been worked out many years ago. Water had accumulated in the abandoned mine, causing large amounts of gas to form. Company officials stoutly denied this allegation, claiming that the two mines had long ago been separated from each other.

Well before the coroner's jury met, several representatives of the coal company also came to a conclusion. They referred to the record-book kept by the fire-bosses, and discovered that a large section of the mine had been fenced off that very morning as unsafe, because of a danger of rock falls and gas. Yet, several bodies were found two hundred feet inside the forbidden area. This was conclusive proof, said the "experts," that the Pittsburgh Coal Company was faultless.

Also, a few bodies were discovered in a section of the mine known as "the gobs." These were worked out areas where water constantly spread out over the floor; hence the "gobs" were always more gaseous than the regular areas. (In later times, "the gobs" have also referred to waste matter brought from the mine and stored on the outside. Thus, a slag heap could be called "the gobs"). The "gobs" (at least those of olden days) provided the major relief places for the miners. Sometimes, to find a dry spot before yielding to nature's call, a miner would strike a match to see where he was. Or, once started, he would add to the comfort of the occasion by lighting up his pipe or cigarette. The example of Naomi a few days back seemed not to have made a single impression. This theory of explosion, plus all the others, were ultimately looked into by the coroner's jury, which slowly began to form.

The President of the Pittsburgh Coal Company, M.H. Taylor, was attending an operator's meeting at Indianapolis' Claypool Hotel when he received word of the Darr tragedy. He held the telegram in his shaking hand, looked his colleagues straight in the eyes, and responded in glorious understatement:

"My God, boys, this is awful!"

John Mitchell, President of the United Mine Workers, gave this statement about Darr:

> There isn't much that can be said, except to call attention to President Roosevelt's recommendation in his last message to Congress that a Commission be appointed to examine into the causes of these

explosions, and make recommendations that will forever end them. Since he has made his recommendations, and in terrible emphasis of his proposition, there have been four horrible mine disasters. Surely no further argument is needed that something should be done.

Rescue work (aided, incidentally, by the lone survivor, pumpman Joseph Mapleton who, his cuts and bruises dressed, joined the teams going back into the mines), was facilitated greatly by the full restoration of the ventilating systems, dispersing the blackdamp. Contrary to several company spokesmen, the mine had been completely destroyed by the explosion. After bratticing in several strategic spots, the rescuers moved apace with their work. As they came to bodies, the rescuers made no effort to haul them to the surface; instead, they kept pushing toward the back of the mine, hoping to find some survivors. Unfortunately, the deeper recesses were in worse condition than areas near the surface. At 5,000 feet, physical damage to the mine was great, and mutilations of bodies was extensive.

For example, Andy Koslaski's right arm was blown off at the elbow, revealing shards of shattered bone, and his head was cruelly crushed. Some victims found in a wrecked car had particles of their clothing blown through the wood of the car and embedded into the coal seam. Two bodies were found at the "swamp base" on the right of the main entry. They were inmeshed into each other so firmly that when the first body was lifted, it was found that his suspenders were wrapped tightly around the other's neck. The explosion ripped along the mine's roof with such force that men in an upright position had the tops of their heads blown away where they stood. Consequently, several headless and limbless bodies were found; so many in fact that numerous "rookie" rescuers were overcome with shock and grief, and had to be taken to the surface.

The coroner of Westmoreland County, Dr. A.C. Wynn, worked out some firm and decisive methods for bringing victims to the surface, and for identification once they reached the top. He ordered that a detailed sketch be made of the exact location of each victim as he was found. Once outside, the man's clothing had to be kept on him and meticulous notes taken on distinguishing marks still visible, and any other things that might assist the authorities and families to identify them. This was a much slower procedure than at the previous mine explosions, and ultimately caused much protest from the townspeople. Nevertheless, the percentage of identification at Darr was higher than at Monongah.

The accumulation of undisposed bodies presented another real problem, one for which Coroner Wynn received much of the blame. By the 21st, for example, most of the Darr victims had been discovered, but only twelve had been brought to the surface, and none were yet buried. Wynn responded to criticisms of his deliberateness by asserting: "I intend that the people shall find out if possible the exact cause of this terrible death bearing catastrophe, and nothing will be suppressed. The [coroner's] jury [which had not yet been formed] will view the bodies and then release them for burials...."

As time ground on, the stench of death emanated from the Darr mouth, enveloping everything within a radius of one mile. Many young rescuers faltered because of the hideous odor, and had to be released from duty. As the Greensburg *Daily Tribune* reported, "Others, used to such scenes and smells, braved it out. But there was not a man among the bravest who did not quelch. By the greatest efforts the bodies were tenderly picked up, but it is being proven imprudent to expose [them] to the air."

"Even horses this morning," said the *Greensburg Press,* "refused to go to the mine. When whipped, they dropped to the ground, and could not be forced to enter. The horses are needed by the rescuers to haul the bodies to the surface, and many of the men say they cannot continue the work without the help of the animals." One young mule, not willing to go into the fearful entry, dropped to its knees after a severe beating from its master, and then arose and savagely attacked the driver, inflicting painful injuries upon him.

All through the days and nights, the harried authorities were accosted by ever-growing emotionally wrought people: a woman in broken English screams out that her husband has not been home, an old man inquires about a beloved son, and several business owners want to know if certain of their customers have perished (leaving one to wonder about the origins of their curiosity). The most persistent question, however, is: "when are decent burials going to take place for our lost husbands, fathers, and sons? It is vulgar to postpone funerals in the name of complete positive identifications."

There were some occasions when Wynn's system worked. Saul Hoke was brought out on the 20th, the day after the explosion. He could be identified only by the clothing he wore. His features were not recognizable, but his clothes were. "That's Saul's shirt," Mrs. Hoke sobbed as she viewed the mangled body. "I washed it on Monday morning." Later in the day on the 20th, Arthur, their fourteen-year-old son, was found by the rescuing party. Joe Simko lay in the temporary morgue on the Darr grounds for two days, friends and relatives passing him by. Finally, a friend told the attending physician that there was one way to identify Simko. If a piece of cotton were found between the toes of one of the corpses' remaining legs, that would be Joe Simko. The burial wrappings were quickly undone, and the piece of cotton found. Simko's wife, as it turned out, had long practiced this procedure so that his remains could be identified if he were killed. On Christmas Day, a headless body was found. The limbs were so mutilated that recognition by this means was impossible. Around his waist, however, was a Knight of Pythias belt. Someone then remembered that the man, while not a member of the Order, had worn this belt on the 19th, in the place of another. He was a foreigner from Jacob's Creek. These, and other examples, seemed to justify Coroner Wynn's slow processing of the Darr bodies.

In the meantime, the crowds swayed back and forth in freezing, ankle-deep sludge caused by an icy drizzle that began late in the afternoon of the 19th. At the mine itself, orderliness in general prevailed, largely because of the confined area. Across the river at Jacob's Creek, however, the populace was constantly in a near-riot condition.

Each time the sky-ferry arrived, for example, numerous scuffles and brawls erupted as dozens tried to get on the little six-seater. Not until sizeable police forces from Smithton and other near-by towns arrived, was any semblance of order restored. Forty-eight-year-old Conrad Schuth of Jacob's Creek found his own peace and order. Crazed by the death of his son in the mine, Schuth flung himself into the Youghiougheny, and ended his sadness. He was a widower, and left four small children. Mrs. Carino Petrano finally found space in the sky-ferry. Halfway across, she climbed suddenly onto its rim and started to jump into the turbulent waters below. Restrained by three strong men, she kept screaming, "Let me go! Let me go! I want to die!" She had lost her husband in the explosion.

Besides the melee at the sky-ferry, many foreigners who had escaped death because of their holiday, continued to celebrate: some because of the joy of still being alive; others to mourn the loss of friends and relatives. One newspaper reported that "all night [on the 19th] there was carousing and disorder, bordering at times on rioting." The foreigners' behavior created additional ill-will toward them by the Americans. The latter were already inexplicably resentful that most of the foreigners had stayed home on the 19th; and, despite this fact, still in some way held them responsible for the catastrophe. The tragedy at Jacob's Creek, just as the one at Monongah, brought out all the biases and prejudices of the local populations. This created conditions that greatly worried the authorities, causing Pennsylvania Governor Edwin Stuart seriously to think about mobilizing the National Guard. The stench that spread from the decaying bodies, however, and the continued rain-fall, dispersed many of the rowdy elements, and figured prominently in the avoidance of riots.

There was one example, however, of how heavy drinking in the context of the disaster, created good-will. The *Greensburg Press,* December 24, headlined this story: "A Quart of Whiskey Saved Six Men from Death." The captions below read: "And it was cheap whiskey at that—a 50 cent grade that stirs murder in the heart of him that drinks it." It seems that two Darr miners had been in Smithton, and split the quart of whiskey between them. When they finished with it, they quarreled, and began to slug each other with fists and rocks, and then there was the flash of knives. Constable J.V. Johnson stopped the brawl, and placed both men under arrest. Their hearing was set for 11 on Thursday morning, the 19th. As witnesses to the fight, four of their fellow miners were ordered to appear in court, furious that they had to lose a day's pay because of some stupid frivolity. As the hearing got underway, the building trembled, but the Justice of the Peace continued with the business at hand. Just as he was about to assess each prisoner with a fine for public brawling and disturbing the peace, an excited messenger arrived with the news that the Darr mine had blown up, and that there was little hope for survivors. "The prisoners who had been eyeing one another with looks of hate sat for a moment, stunned by the awful news. Then realizing by what mere chance they had been saved, they clasped hands, and tears trickled down their faces."

Mrs. John Neidermeir stood with her three small, wailing children at a telephone booth in Jacob's Creek most of the day on the 19th. She wanted to make a long distance call (rare enough in those days from a public pay phone) and tell Teresa, her sister in Pittsburgh, about the great calamity. John was in the Darr mine, broken and shattered, and already decomposing—leaving his wife and children behind. Mrs. Neidermeir explained in broken English to a reporter from the Pittsburgh *Gazette-Times*: "We have lived here for four years, and John has worked in the mine. I have always dreaded something like this. When he said goodbye this morning, something seemed to tell me it was for the last time. All morning I thought of him—but I busy myself with the children and the housework. When the sound of something like an earthquake come, I knew what had happened at once. When friends came to tell me that there was still hope that he might be taken out alive, I know it could not be. Tonight, I am so lonely...." Throughout the entire area of Jacob's Creek, variants of Mrs. Neidermeir's story were repeated by scores of widows and other survivors. There had been presentiments for them, too, when their men went into the mine that morning; likewise, many, many things had suddenly become clear for them—but now all in retrospect.

In addition to playing havoc with the economic and social life of Jacob's Creek, the great mine explosion also destroyed much of its political capabilities and activities. As Amiel Ginter, Chairman of the Board of Elections, put it: "The majority of the...Americans known to be lost, lived on this [Jacob's Creek] side of the river. All owned homes, and were excellent citizens. They were a fine lot of boys, and did their part well in making this a prosperous little town. A majority were young married men, and paid for their little homes. It is mighty hard on those left behind. Yes, we got out 91 votes at the last election and 69 were Republican. We can't do it the next time, as [most of]...those boys have been lost. The Democrats have the best of it now."

And, so, on and on the stories went. Of how a ten dollar bill was found wrapped in a piece of cloth around a shattered, disconnected knee. Of how a foot was found sticking to the roof of a passageway. Of how at least 325 men went into that mine this morning, according to Mr. Michael Hallapy, an organizer for the Miner's Union, and yet the coal company says that the number of casualties will be below 200. If this is so, why have 300 coffins been ordered? Of how forty-five houses in a row in Jacob's Creek had funeral wreaths pinned to their front doors. Of how very little cash was found on the bodies of the foreigners, because they "always horde their money in trunks and valises," and of how in some way this fact constituted yet another reason for anger and resentment against them. It was as if the Americans were saying to the foreigners: "Why couldn't you have died instead of us? Why? Don't you see that our lives are more valuable than yours?" Of how four hundred foreigners left two days after the explosion to go back to the Old Country (if they could afford it) or to Patterson and Passaic, New Jersey, there to work in the textile mills. "Good riddance," exclaimed some of the locals; "let them go back to Italy, Poland, and Hungary, and bring bad luck to someone besides coal miners for a change." And of how this would indeed be a bleak Christmas in the coal mining regions of Pennsylvania and West Virginia, to say nothing of Alabama.

Amid the ever-growing stories and rumors, Coroner Wynn's steady work continued of retrieving and identifying the victims. He had them brought to an extension of the blacksmith shop, hastily built to serve as a temporary morgue. By Christmas Eve, Wynn was ready to release the bodies for funerals and, as it turned out, seventy-three of these went to their graves unidentified. Still, this was a lower percentage of unknowns than at Monongah.

Funerals from Connellsville to Smithton marked the celebration of Christ's birth in those closing days of 1907. Even hardened members of the State Constabulary and of the Press Corps were seen frequently to turn their heads sharply from these scenes, so that no-one would see their tears. The co-mingling of life and death was symbolized by Father Adam Bint of St. Timothy's Church in Jacob's Creek when on Christmas Day he baptised ten orphaned infants, and then said requiem mass for their lost fathers. Father Bint's immediate predecessor at St. Timothy's, the Reverend Father Lawrence A. Carroll, now of Kittanning, was also on hand to comfort the bereaved. But even these gentle men of the cloth could not ease the pain and hurt that permeated the little valley on that sad Christmas Day.

Most of the victims were buried near Smithton, in a special

cemetery plot acquired by the Company. This seemed to be a more feasible plan than trying to transport bodies across the river. It was broadcast throughout the nation that the Pittsburgh Coal Company was paying all the burial expenses for the dead miners, a report that earned for it much goodwill. Then it was discovered that each miner had been forced, as a condition of employment, to take out insurance for himself to the amount of $150.00. The coal company deducted burial costs from this figure, giving the remainder to the beneficiary. There were also rumors that the coal company directly sponsored various benefit performances to collect money for the survivors, as well as direct newspaper appeals for the same purpose: all in an effort to lessen its own financial loss as much as possible. The attitude of the Pittsburgh Coal Company definitely was not as benevolent as Watson and Crowe had been in West Virginia and Alabama. The reasons for this, according to newspaper reports, were the company's bad financial condition brought on by poor management, and its adamant refusal to have anything to do with labor unions, particularly the UMWA. The company's attitude and actions produced an enduring bitterness.

Immediate relief for the Darr sufferers was provided by contributions from several near-by fraternal organizations. The Elks Lodge of McKeesport, for example, sent a car-load of clothing, toys, and candles. The coal operators allowed the families to continue living in company houses, at least through the holiday period. Small amounts of cash were locally collected, but no effort was made until January, 1908, to form a National Relief Committee. The reasons for this delay were fairly obvious: U.S. citizens were still being asked to help the Monongah victims, and they now had large, personal expenditures because of the Christmas season. Also, the Red Cross was wary, remembering all too well Miss Byington's difficulties at Monongah. Joseph Steinmetz, Pennsylvania Branch of the Red Cross, wrote to Logan McKee, Secretary of the Pittsburgh Chamber of Commerce: "We still believe a Red Cross Committee in Pittsburgh could operate to an advantage, as in that way they could reach the entire United States through the national body.... A prominent organization with such excellent control, and removed of petty jealousies of local influence, would be exceedingly beneficial for relief work in times of need. We will not do anything in the matter at Darr until we hear from you. We feel that you are in the field, and we have no right other than through you and by your cooperation." This demeaning attitude assured that the Red Cross would play only a minimum role at Darr.

On January 2, 1908, the Darr Relief Committee was formed, with Governor Stuart and Pittsburgh Mayor, George W. Guthrie, the Honorary Chairmen. Also serving on this Committee were the mayors of McKeesport, Connellsville, and Smithton, as well as the Consulate Generals of Austria-Hungary and Italy, both at that time residing in Pittsburgh. An immediate appeal went throughout the country for assistance to the 542 widows, children, and other dependents of the Jacob's Creek disaster.

The Darr Committee provoked criticism by professional welfare workers around the country because of some of its practices. Edward T. Devine of the Charity Organization Society of New York wrote, January 8, 1908, to Boardman of the Red Cross:

> The unfortunate feature of the whole thing appears to be that because the company has not treated families quite as well as at Monongah, the general public is also in danger of treating them badly. In other words, the families who were not at fault, seem to be in grievous danger of suffering serious neglect.... I do not wish to be considered as criticizing the Constitution of the Darr Mine Relief Committee, which includes prominent and responsible men. Mr. Wilmot [of the Carnegie Hero Fund] considered it unfortunate, [however], that it should include the consuls of Austria-Hungary and Italy who, from the nature of their positions, will almost inevitably become in a degree the representatives respectively of the Austro-Hungarians and the Italian claimants of the fund.

Devine went on to deprecate the presence of labor leaders on the Committee, fearing that they would favor the survivors of union workers over those of non-union members. No evidence exists, however, to show that this actually happened. On the contrary, the Committee was chided for its decision to treat all cases alike, "corresponding to an insurance board," and not acting upon welfare principles of individual need. J. Bryon Deacon, a professional welfare worker, criticized the Darr

Committee "for being too concerned with escaping the arduous and exacting labor which a program of truly individualized treatment requires." He conceded, however, that though there were no trained social workers on the Committee who were responsible for disbursing the funds, "the distribution was surprisingly intelligent."

As at Monongah, the survivors were personally investigated to determine past economic and social habits. Whether adverse or not, these factors, also as at Monongah, did not seem to weigh heavily in terms of final reckonings. The Darr Committee, like its West Virginia counterpart, made no distinction between natives and foreigners living here or abroad, in determining relief measures.

One practice of the Darr Committee had a permanent influence on the social legislation of the time. As Mabel Elliott explained: "A unique aspect of the benefits was the plan for giving to certain widows and other dependents deferred payments to protect them from unwise expenditures and to safeguard their money for future needs." For those opting for a monthly allowance, each widow could receive $15.00, with an additional $5.00 going to each child under sixteen. Other relatives and dependents were eligible for $10.00 a month, presumably like the widows and minor children, for life, unless their statuses changed. The Darr practice of deferred payments was later refined by the Red Cross after the Cherry Mine explosion in Illinois; still later, when the federal government adopted pension procedures for disaster and social security purposes, it followed closely the pioneering efforts of the Darr Committee.

The deferred payment play of the Darr Committee differed from the Monongah Committee's practice of giving advances against a final settlement. The Darr survivors, unlike those in West Virginia, had a choice between a lump-sum settlement, and an on-going pension. By mid 1909, all settlements had been made with the Monongah survivors; yet, as long as Darr beneficiaries remained widows or minor children, they received money from the Relief Committee. Thus was produced yet another example of how the private sector in the United States paved the way for welfare and pension programs which, in all too many cases, the government received the credit for "discovering."

No amount of "deferred payment" plans, however, or

Cherry Hill, Illinois explosion; November, 1909. Photo courtesy, UMWA.

sympathy from the public, could assuage the sorrow that kept hanging like a shroud over that little mining community during the Christmas season of 1907. By the first of the year, the coal company had torn down all the structures that had been built as temporary morgues, and it would not be too many days before the Darr pits were re-opened, with scores of men going into them each morning. Thus was erased all the physical evidence of the great catastrophe that came to Jacob's Creek on the 19th. The mental scars, of course, remained, inflicting upon an entire generation thousands of sleepless nights, wherein the remembrances of things past became the all embracing stimuli to their very existence.

Time does not wait for human emotions to catch up with it. Inexorably and inscrutably, it brings an event upon us before we are able and willing to accept and assimilate those that have gone before. The Naomi explosion occurred on December 1; a scant six days later, the Monongah horror unfolded. Citizens around the country still had Monongah on their minds and lips ten days after when Yolande erupted. Then only thirty-six hours later, on the 19th, came the holocaust of Jacob's Creek. What would happen next? became almost a universal question. When a week had passed after Darr, and no other explosion had occurred, the nation was relieved. On the tenth day, there was almost cause for rejoicing. But then, there came the twelfth day after Darr—December 31. Just as the first day of the dreadful month brought harbingers of death, so it was with the last.

Death's grudging farewell to the coal miners in 1907 came in the magnificent Southwest, long celebrated in song and story by the American people. The great territory of New Mexico, wrested from the Mexicans in 1848 by the Treaty of Guadaloupe Hidalgo, and to become our forty-seventh state in 1912, hosted the final scene of the dreadful month. The territory's first coal mine, operated as early as the Civil War, was located near Carthage, and was known generally as the Government Mine. It was worked by union soldiers to provide the smithing needs of Forts Selden, Bayard, and Stanton. Twenty years after the war, in 1881, the Carthage Mine made news when two army six-mule teams hauling its coal were caught in the sands of the Rio Grande River, and lost. Despite this set-back, the mine continued to attract workers, and by early 1907, Carthage's population was over a thousand.

Though there were some Europeans on the scene, as could be expected, most Carthaginians were Mexican.

The accident occurred in the Bernal Mine, owned by the Carthage Fuel Company. Approximately sixty men were employed by the Company, but for various reasons, only forty had gone into the mine on Tuesday, the 31st. At about noon most of that number, fortunately, came to the surface to eat their mid-day meal. After just a few bites from their sandwiches, the miners felt the ground shake and tremble beneath them. As in Pennsylvania, West Virginia, and Alabama, they did not need to be told the origin of the turbulence.

W.L. Weber, the Bernal Superintendent, was instantly on the scene, to direct rescue operations. On hand, too, were the community's women, ready and willing to offer their services in caring for the dead and wounded. As the *Socorro Chieftan* reported:

> On every hand there were abundant evidences of the terrific force of the explosion. The bodies of some of the dead were blown against the walls of the mine with such great force as to flatten them almost beyond recognition. In one instance at least, a dead body was identified only by parts of clothing adhering to mangled flesh. The body of one miner who was coming out of the mouth of the mine when the explosion occurred, was shot a hundred yards into the air as from the mouth of a cannon, and nearly all the bones were broken by the fall.

There had been fourteen men remaining in the Bernal Mine at noontime, and it was later learned that eight of them died instantly. Of the six left, three succumbed, bringing the number of fatalities to eleven for Carthage, and 702 for the nation.

Several of the dead and injured in the Carthage explosion had small stones driven into their flesh like bullets, requiring Dr. H. Bacon, the company surgeon, to work feverishly to remove them. Many of the miners expired while he was performing these operations.

The mine itself was not heavily damaged by the blast, and could have been re-opened immediately if the miners had so chosen. It closed on New Year's Day, however, for the funerals, and started up again on January 2, 1908. The miners thought work was the best means of relieving the nervous tension and strain under which they had been operating since the blast. Almost as quickly as the miners' return to Bernal was the verdict of the coroner's jury. It was declared that a "windy shot" was the culprit, a shot that raised huge amounts of combustible coal dust which was ignited by another shot going off at the same time. That is why, said the jury, C.L. Wilcox, Giovanni Andeole, Joseph Canya, Lorenzo Roverto, and seven of their fellow workers met such a fearful death on that last day of 1907.

And so time, benevolent time, brought the month to a close. Officially, 702 men lay dead from coal mine explosions during December, and the Lord only knows how many in addition to that number were also lost. Were these statistics not sufficient evidence, asked the reformers, that something be done? Standards of quality, perhaps, in explosive powders? A strict division of labor between digging the coal and firing it from the walls? Wasn't it time to hire specialists for the latter activity? More inspectors to see that bosses and workmen properly did their jobs? Steps to assure proper ventilation, and to make sure that mines were not connected to each other in the innermost bowels of the earth?

These questions were only the start. Dozens more raised themselves as "experts" and citizens alike tried to fathom the meaning of this dreadful month of December, 1907. And with every question, every suggestion for reform in coal mines and other areas of American industrialization, the wealthy and the powerful raised the clamor:

> "But, my God, don't you know those things are socialistic?"

Aye, there was the rub: "socialistic," always a word in America of extremely negative connotations. That is what held back reform for so long, making us the last great industrial country, after Germany and England, to do something for the lower rungs of capitalism.

But, then, the time came in the United States when there was a difference between "reform" and "socialism," a difference brought on in considerable part by journalists and by President Theodore Roosevelt. Then, too, all the coroner's reports dealing with the events of December, caused thoughtful people of good will to realize that now was the time for the government to prove its claim of late, that it was the "first servant" of the American people.

Sources

Annual Report of the Pennsylvania Department of Mines, 1907.

Birmingham *News,* December 16, 18, 19, 20, 1907.

Elliott, Mabel A. "The History of the American National Red Cross," Volume 20-A. *The American National Red Cross,* Washington, D.C., 1950.

Fairmont *Times,* December 19, 1907.

Greensburg (Pa.) *Daily Tribune,* December 21, 23, 24, 26, 1907.

Greensburg Press, December 20, 24, 1907.

MSS. American National Red Cross, Washington, D.C.

Montgomery (Ala.) *Advertiser,* March 31, 1910; May 26, 1932.

Montgomery (Ala.) *Advertiser-Alabama Journal,* January 18, 1938.

Pittsburgh *Dispatch,* December 18, 19, 1907.

Pittsburgh *Gazette-Times,* December 20, 21, 22, 1907.

Pittsburgh *Press,* December 20, 24, 1907.

Sherman, James and Barbara. *Ghost Towns and Mining Camps of New Mexico.* Norman: University of Oklahoma Press, 1975.

The Socorro Chieftan (New Mexico), January 4, 11, 1908.

Perspectives II

Thirties, Forties, Fifties

During the 1930s, coal mining fatalities dropped significantly from the preceding decade. As is well known, the twenties were "boom," while the thirties were "bust." This economic difference explains best why 2,369 miners were lost to explosions in the twenties, compared to 696 in the thirties. The saving of lives in the latter period must be attributed primarily to economic factors rather than an over-riding rush of humanitarianism.

Possibly the most astounding thing about coal mine explosions of the thirties was that West Virginia contributed less than a hundred victims. Ohio and Illinois now became the scenes of the major catastrophes.

Millfield, Ohio was a typical village in the Hocking Valley. Its mine was owned by the Sunday Creek Coal Company, the largest in eastern Ohio, with W.E. Tytus as President. Three hundred men were in the mine on the morning of November 5, 1930, when it exploded. Most of the miners emerged safely; 82 did not.

President Tytus and several company officers were escorting a group of Zanesville and other area businessmen through the mine when the explosion occurred. Most of these men, including Tytus, became victims.

The businessmen were inspecting several old sections of the mine, with a view to buying them and resuming production. These sections had been bratticed off from the main working areas. As the business party had an entry knocked into one of these old sections, a "squeeze" occurred. This is coal-mining jargon for a sudden collapse of the roof and walls in a section of a mine. It was speculated by many that the "squeeze" released deadly gases into the mine proper, and these were, in some unknown way, detonated.

Other explanations of the explosion placed its origin in a new air shaft under construction. The fan that was being fitted

110

into this air shaft, however, was not damaged. It appears likely that the main explosion started from the "squeeze."

Rescue operations were seriously hindered by the blackdamp pouring out of the mine's mouth, and by the 8,000 spectators who arrived quickly as news of the disaster spread. (To get a good view, several people climbed onto a platform next to the mine shaft. A rescuer shouted to them that they were standing on an old board a few inches thick which covered a hole 190 feet deep. Naturally, the crowd quickly dispersed).

The rescuers noted that more men died of gas than from the explosion itself. As one survivor, George Raft, related, "I ran along the passageway toward the main shaft. I heard cries of men behind me. They were loud and sharp at first, and then choked." The position of numerous bodies showed that they had made desperate attempts to escape the gases and to reach fresh air. Elmer Dingledey, reporting for *The Athens Messenger,* went down into the mine with a team of rescuers. He gave this vivid picture of what happened:

> "God, there's Jim, my buddie!" "Lord, Pa, you ain't dead?" were reverent cries that we heard. There in the smoky, murky glimmer of flashlights, tears were shed by calloused, bull-shouldered men, miners all of them. But the work went ahead."

By far the most poignant coal mine explosion of the thirties was at Moweaqua, Illinois. It occurred on Christmas Eve, 1932, killing all but two of the 56 man day shift.

The Moweaqua mine operated as a symbol of hard times. Glen Shafer of Pana, Illinois, owned it, but had closed it a half-year before because it was unprofitable. He then leased the mine to a Citizen's Committee in Moweaqua which, in turn, gave employment to 104 miners. Each of these miners was a stockholder in the mine; however, they barely scratched out enough wages each week to provide for their families. But, at least, it was a "living."

The explosion was announced on the outside by severe rumblings of the ground. Rescuers found the origins of the trouble some 1,700 feet back. Seals of canvas and other materials had been hung along the main haulways to contain and control methane gas (although, according to State Mine Inspectors reports, the mine was "non-gaseous). The seals were

Crowds waiting at an un-named mine disaster; ca. 1920s. Photo courtesy, UMWA.

so poorly built that any sudden change in pressure (easy enough in the cold month of December) would send clouds of gas out into the main entries. This is undoubtedly what happened after an extensive fall of dirt. A miner's open light came into contact with these gases.

The screaming children and weeping wives of the entombed men stood in a cold down-pour at the mine's entrance on that Christmas Eve of 1932. To be sure, all the Season's festivities were cancelled because of the blast, and the big Christmas Tree in the town square was taken down. Besides, Tom Jackson, the town's Santa Claus for many years, was lost in that dreadful explosion. In the local newspaper, the company carried this notice:

"The Moweaqua Coal Company has COAL at the mine and can take care of our customers. Buy from us and help toward reopening the mine."

And so it went.

Toward the end of the decade, the definite economic upturns were steadily solving the nation's unemployment problems. The prospects of a drawn-out war in Europe also heightened this feeling of industrial optimism.

Along with the increased activity, however, came a resumption of the negligence which had characterized coal mines for so long. In fact, the first three mine explosions of 1940 were "major," that is, each of them took fifty or more lives.

Everybody knew that West Virginia could not hold out forever, and on January 10, 1940, the Pond Creek Mine at Bartley blew up, taking with it 91 victims. The top officials of the Pocohontas Coal Corporation, owner of Pond Creek, were holding a safety conference on the mine grounds at the very moment of the explosion.

The Pond Creek Mine was known to be very gaseous, and also one in which deep deposits of coal dust had accumulated. It was the same old story: the gas was ignited by either an open light or an electric arc, and then propagated through the mine by the dust. The explosion gave the attendants at the "safety conference" plenty to do and think about for the next several days.

The next explosion of 1940, on St. Patrick's Eve, was in Neffs, Ohio, and it took 72 lives. This mine was termed by owner Hanna Coal Company, as its "safest," and the "most modern" in Ohio, equipped with the "latest air-conditioning safe-guards."

In 1936, First Lady Eleanor Roosevelt had toured the mine at Neffs, and had exclaimed, "It's too clean to be a mine." When she heard of the explosion in 1940, she remarked, "I thought that was a model mine. Things seemed all right to me there in 1936. But I'm no mine expert." Perhaps Mrs. Roosevelt did not know that just about every mine in the country that had ever blown up was a "model mine," at least according to company officials and state politicians.

The over-use of "non-permissible" black powder seems to have been the culprit in this explosion. Though the mine was classified as "non-gassy," the explosion was propelled by gas and un-watered coal dust.

Pennsylvania was again visited by death and destruction on July 15, 1940, when an explosion swept through the Sonman E. Mine at Portage. L.C. Campbell, Vice-Presidential assistant of the Koppers Company, owner of the mine, asserted shortly after the explosion that the number of fatalities could not possibly go above 47. The final tally was 63.

Edward Ben, a survivor, "saw a flash and heard a noise." Then, he said, "things started to fall all around me. It was difficult to breathe and we got down on our stomachs for a while. The air was better down there."

Methane gas released by a rock fall got into a haulage way where it was ignited by an electrical arc from a locomotive. Fortunately, the explosion was not strong enough to stir up the coal dust, and thus be propagated throughout the rest of the mine. If it had been, the tragedy would have been much greater than it was: there were 350 men in that mine at the time of the explosion.

And now we move back to—where else?—West Virginia, where on May 12, 1942, another of its mines claimed 56 lives. This was the Christopher Mine at Osage, and the explosion was caused by an electrical arc igniting un-watered coal dust.

One would think that a West Virginian could forego the temptation to be a spectator at a mine accident. Yet, for several hours officials and ambulances could hardly get through because of the "hundreds of cars that choked State Route Seven and the narrow cinder road leading down a hill to the operation." And so it went.

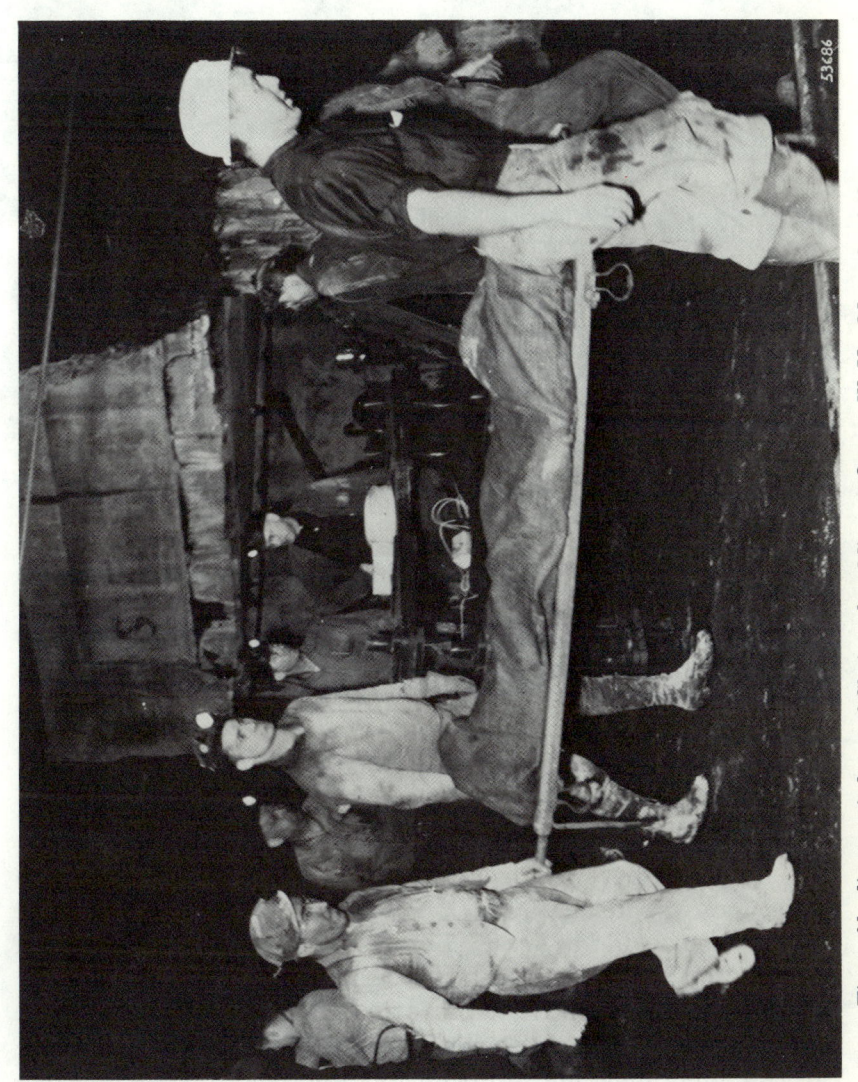

First of bodies carried out of Christopher Mine at Osage, W. Va. May 12, 1942.

Scene at Christopher Mine at Osage, W. Va. May 12, 1942.

The next major explosion was in Washoe, Montana, on February 28, 1943. This one killed 74 men. It was caused by gas coming into contact with an open light. Though the mine was gaseous, the bosses still permitted the workers to smoke.

The last two explosions to be mentioned in this perspectives chapter were Centralia and West Frankfort, both in Illinois. They were tragedies of great magnitude, and they attracted world-wide attention. They also caused the federal congress to do something about safety in the coal mines.

The weather was freezing in Centralia, Illinois at mid-afternoon on March 25, 1947. In its mine, 142 men labored; soon 111 of them would be dead. The mine was "exceedingly dry and dusty, and heavy deposits of coal dust were present along the roadways and on the roof. . . and timbers in working places and entries." Shot firers set to work about 3:30 p.m. Apparently one of the shots blew out, igniting gas and coal dust. When one considers that blown-out shots were major causes for explosions at least as far back as 1900, one may conclude that in 50 years, the coal industry had not learned very much.

This mine had been inspected just one week before the explosion. Driscoll O. Scanlon, Illinois State Mine Inspector, gave this report:

. Ventilation is inadequate in several areas of the mine.

. The condition of haulage roads is bad. They are dirty, dry, and dusty. There is loose roof on the haulage roads in a number of places.

. A wall in the underground machine shop is broken and bulging out into the room.

A letter from an anonymous group of Centralia miners, March 3, 1946, to Illinois Governor Dwight H. Green, however, showed the small influence that Scanlon had: "Mr. Scanlon is the best inspector that ever came to our mine," but the Mining Board would not let him enforce the law—primarily because of political and economic considerations. The letter closed, "Please save our lives." Of course, the letter was ignored.

Rescue operations were slowed down by the hundreds of cars and thousands of spectators at the site. Also, since copious quantities of gas were still present, another explosion was feared; therefore, motorized cars could not be used by the rescuers. Instead, they employed mule-power, which meant that it would take a week or more to get to the stricken miners. The first miner's body to be recovered was that of 74 year old Mark Watson of Centralia. He had been a miner for more than half a century. In Watson's and his 110 buddies' honor, UMW President John L. Lewis called America's half-million miners off their jobs for six days in a "memorial strike."

One Centralia rescuer bitterly remarked:

> "Some of the people who have been giving the coal miners Hell should wake up now. I'd like to have those guys here...the ones who are always yapping about the miners being over-paid. I suppose those guys down there are over-paid. Yeah, they're over-paid, and this is the pay-off. I can think of some guys I'd like to grab by the neck and take down that shaft with me." Another rescuer said, "I used to joke with the boys about how deep the coal dust was...but I guess we never took the situation seriously enough."

Like most bituminous mines at the time, Centralia was under the control of the federal government. After a series of strikes in the Spring of 1946, President Harry S. Truman issued Executive Order 9728, nationalizing the coal mines. This fact of governmental control at the time of the Centralia explosion gave many members of Congress a chance to indulge in their favorite pasttime of grandstanding. The Republicans cried "bureaucracy," and the Democrats yelled "politics," as the solons generated hot air out of Washington.

Republican Senator Styles Bridges of New Hampshire used the occasion to blast Interior Secretary J.A. Krug (under whose jurisdiction the federalized coal mines came). The Secretary was, said the Senator, an "arrogant bureaucrat," who was more interested in campaigning for the vice-presidency of the United States than in enforcing safety codes in coal mines. (As usual, John L. Lewis was more to the point: he said Krug "murdered" the Centralia miners). Another Republican, Homer E. Capehart, of Indiana, said the Centralia tragedy was "proof that the Federal Government should not try to run industry." (Just what this statement meant in reference to 111 men lying dead was not made clear by the Senator. Surely he did not mean that the men died *only* because of federal control. There were far too many precedents under "private" management for the veracity of any such claim).

For their part, the Democrats downplayed these Republican offenses. Glen Taylor of Idaho, said that Republican complaints were "political drivel," because it was well known" that the Government had only run up the flag over the mines, but had left the running of them to the operators. Alben W. Barkley of Kentucky, wanted to make sure that any subsequent congressional investigation of Centralia cover *all* officials—federal, state, and local—rather than spotlight the Department of Interior. He said that a disastrous mine explosion occurred the year before in Kentucky "without any great hub-bub about it," and the federal government was not operating the mines then. (He was apparently referring to the explosion of the Belva Mine at Fourmile, Kentucky, which killed 25 men. It happened on December 26, 1945.)

The Senate approved *viva voce* a resolution submitted by C. Wayland Brooks, Republican of Illinois, to investigate the disaster at Centralia. It appropriated all of $5,000 to the Public Lands Committee for this purpose.

A federal mine safety code had been adopted on July 24, 1946, shortly after the government take-over of bituminous mines. Not much, really, was ever done to enforce this code. Navy Captain C.H. Collison was the federal official responsible here. He said that because of a lack of personnel, his office did not check mines immediately to see if they had followed safety inspections. (Robert Medill, Illinois director of Mines and Minerals, was ostensibly the state man in charge of safety. He had rejected a request two years before to close Centralia because it was unsafe. He also had reportedly ordered that any safety notices posted on the mine's bulletin board be quite brief, on the grounds that lengthy pieces of writing would scare the non-English speaking miners. After the explosion, Medill resigned for, as Governor Green put it, "the good of the service.")

The Centralia blast changed some safety procedures. The Bureau of Mines (which only had the power of recommendation—not enforcement) reported to the Coal Mine Administration that 518 mines under federal jurisdiction had unsafe practices, and were running a grave risk of explosion. The federal agency closed down these mines forthwith, an action that the Bureau of Mines had never been allowed to take.

There followed now a frenzied time in which the operators busily improved their ventilation systems, rock-dusted their mines (rock-dust is not volatile and serves, as does water, as a preventive against clouds of coal-dust being formed during an explosion), and changed their blasting procedures. They did these things primarily because the government would not let them reopen until they had complied with federal directives. Closed mines meant lost production and lost profits, so compliance came rather quickly. But then, on June 30, 1947, governmental control of the mines ended, and along with it, the enforcement of at least a modicum of safety standards. Before long, it was "business as usual." And so it went.

During the next four and a half years, there were plenty of "minor" disasters in America's coal mines: 15 killed here, 20 there, etc. It was not until December 21, 1951, that another "major" explosion happened. It was at West Frankfort, Illinois. Arlie Cook, day manager for the mine's owner, Chicago, Wilmington, and Franklin, predicted that the death toll would be about 90. Deneen Taylor, a pit boss, disagreed, arguing that the fatalities would be greater than Centralia. Unfortunately, he was right: 111 men had died at Centralia; 119 perished at West Frankfort.

The mine, one of the largest in the world, produced between three and four million tons of coal a year. It had more than twelve miles of tunnels, larger in area than Manhattan Island.

A day before the explosion, a certified examiner, James

Wilson, inspected the West Frankfort mine. His finding was:
"NO FAULT."

The mine was gaseous, and quantities of methane were known to exist in certain abandoned sections, called "old ends." In July, 1951, federal inspectors W.R. Chick and C.L. South, had recommended that these areas be sealed; or if not sealed, at least not allow the air piped to them to return through the main haulways. West Frankfort Superintendent, John R. Foster, openly dismissed these recommendations, because, he said, they dealt with "controversial matters."

According to the Bureau of Mines report, "all working sections affected by the explosion were ventilated by air "that had been in the 'old ends.' " Just before the explosion, three of the "old ends" caved in, releasing large amounts of gas, which were blown out by the ventilating system into the main area. There the gas came into contact with an open light. Was it an arc from electrical equipment? Or was it from someone smoking? The Bureau of Mines said it was "probably" an arc, while other investigators found numerous spent matches and cigarette butts, though there was a "strict management ban on smoking."

As in Moweaqua nineteen years earlier, the Christmas Season added to the poignancy of the West Frankfort holocaust. Many of the dead miners had worn their best clothes that night, expecting to change back into them the next morning for Christmas shopping with their families.

Christmas is a time for miracles, and one came to that saddened community on the 24th. Rescuers found Cecil Sanders after he had been entombed for sixty hours. He was dazed and weakened, but he recovered. B.R. Williams, however, was not so fortunate. He wrote to his wife, Laura, on a piece of cigarette paper:

> "I love you all way. I go tonight with Christ.
> I love Him, too."

Fourth grader, Phyllis Jean William, penned this remembrance to her lost father, Carl:

> "The Time has come for us to part along the sunny bank. The time has went by so fast that it gives us no time to think. You was the one that I adorde. And I

love you still.
"If you could only be back on this earth where a will is
a will."

Old skepticisms cropped out after the explosion. Trinity Methodist pastor, H.Y. Sladen, preached: "I don't know what impression the tragedy will make upon the community, but I have the feeling that most of the people who are stirred today will, in a few weeks, have forgotten about it." A local columnist agreed, saying that the explosion "will hold a place on the front page for only a day or two and the public interest will shift to elections and scandals and L'il Abner." Remarkably, however, these skepticisms were not as accurate as they always had been in the past.

The explosion brought several high ranking officials to West Frankfort:

. Illinois Governor Adlai E. Stevenson was there, casting dreadful maledictions on the operators and state legislators for not adopting his recommended safety code. He also promised to deal harshly with any state official found to be negligent in the matter.

. John Forbes, Director of the Bureau of Mines was present, but could only shake his head.

. John L. Lewis was there. He issued a statement, and perhaps his more than usual trenchancy was justified:

> "I only wish those members of Congress who have consistently blocked the enactment of a statute giving the federal Bureau of Mines enforcement powers could be here in person and look into the faces of these terribly mangled mine workers.... I am sure they would forever afterward have trouble living with their own consciences and might be induced to permit coal miners to continue to live while they continue to draw their salaries as members of Congress."
>
> "It is a sad commentary that so many men have to die and so many widows and orphans created before certain types of politicians will ever give consideration to the right of the coal miners to have security while they produce the commodity upon which our economy is based."

By far the greatest miracle to come from the West Frankfort explosion was that it caused the federal congress, for a change, to do something for the common coal miner in the

United States. The Federal Coal-Mine Safety Act was passed in July, 1952, and it gave long-awaited enforcement powers to Bureau of Mines inspectors. If, in the opinion of an inspector, a mine is dangerous, he can order the withdrawal of the regular workers until the problem is corrected. And, of course, there were plenty of operators and politicians around to label this Act "socialistic," and destructive of "states rights."

The law stuck, though, and for the next seventeen years there was not a major mine explosion in the United States, proving to all but the die-hards that it was effective. But, then, on November 20, 1968, there came Farmington—in West Virginia.

Sources

Centralia (Illinois) *Sentinel.* March 24, 27, 28, 29, 1947; April 2, 1947.

Columbus (Ohio) *Evening Dispatch.* November 6, 1930.

Congressional Record. 80th Cong., 1 Sess. Vol. 93, parts two and three. pp. 2612, 3229, 3230.

Humphrey, H.B. *Historical Summary of Coal-Mine Explosions in the United States, 1810-1959.* Washington: Bureau of Mines, 1960.

Louisville *Courier-Journal.* March 17, 1940.

McDonald, H.L. *The Millfield Mine Explosion, November 5, 1930.* The Plains, Ohio: privately published, 1960.

New York *Times.* November 6, 1930; December 25, 1932; January 11, March 17, July 16, 1940; May 13, 1942; February 28, 1943; March 26-7, 1947; December 22-4, 27, 1951.

The Athens (Ohio) *Messenger.* November 5-7, 10, 1930.

The Cleveland Sunday News. November 9, 1930.

The Columbus (Ohio) *Citizen.* November 6, 1930.

The Daily American (West Frankfort, Illinois). December 22, 24, 26-8, 1951.

The Moweaqua (Illinois) *News.* December 30, 1932; January 5, 18, 1933.

Chapter Five

Why?

What caused all the mine disasters in December, 1907? Was it, as some superstitious people believed, a chain reaction in which fate dictated that when one catastrophe occurred, it would necessitate at least two more to appease the gods? Does tragedy simply have to run its course?

Or could it be this? Consider that, excepting Naomi, all the explosions came around mid-morning. Weather conditions on the days of the explosions were roughly the same at Monongah, Yolande, and Jacob's Creek, the places with the highest casualty rates. Now, note that there were certain geographic similarities: they were all about the same altitude above sea level, and in the same longitude. Was it not possible, therefore, for seismic conditions to have been to blame? Yes, of course, replied a sizeable portion of those who were looking for reasons.

Not so, said Dr. J.W. Holland, director of Carnegie's Museum in Pittsburgh, as he rather unsatisfactorily explained: "The fact that the explosions took place in the same zone of altitude and in about the same longitude...proves nothing. The coal mines are at that level in those longitudes. That is all there is to it."

Despite this disavowal, seismic explanations continued to be given through January, 1908, and in the months ahead. As it turned out, this "reason" for the explosions was about as forthright and decisive as those issued by the experts and coroner's juries. It dismayed the laity to witness such frequent and profound disagreements among those who had supposedly been trained to get to the heart of such matters.

As previously noted, the Naomi coroner's jury rendered its decision very quickly: an explosion of undetermined origin killed thirty-four workers. The official decisions for Monongah and Jacob's Creek, however, were harder to come by than at Naomi. For one thing, several jurors complained about being

confined during the Christmas season; thus, the official
Monongah report was not given until January 15, 1908. In
defense of the coroner's jury, it must be pointed out that its job
was stupendous; when its final report was given, it came to
about nine hundred typewritten pages.

While Coroner Amos and his jurors went about their tasks,
speculations and questions in the area remained at a high
level. Here were the most common ones:

. What was the origin of the Monongah explosion? Was it
the runaway trip in Number Six or a blown out shot in Eight?

. Just exactly how many casualties were there?

. Would any steps now be taken, or even recommended, to
prevent future catastrophes of this sort?

West Virginia was well known at that time as a coal-
mining state with careless, even nonchalant, attitudes toward
safety. The consensus about improvements being made,
therefore, was gloomy.

Numerous inspection teams, both private and official,
wandered through the winding passageways of Numbers Six
and Eight, trying to come to their own conclusions about what
caused the great tragedy. Many of the findings of these
inspections made their way into the final coroner's reports.

The team that got the most attention was from France
where, in March, 1906, over a thousand men were lost at the
colliery at Courriers, the greatest coal mining tragedy in
recorded history to that date. Jacques Cafinel and Paul
Dumaine headed the French delegation. After an initial tour of
the mines, Cafinel gave this statement:

> The thing that impressed me the most was the terrific
> heat caused by the explosion. Why, we found
> conditions today that convinced me beyond a doubt
> that the temperature was many times greater than
> that which resulted in the Courriers horror....
> In one place, and in particular in Mine Number Eight
> today, we encountered evidence of a higher
> temperature than I ever believed could be caused by
> an explosion. The Courriers horror was...somewhat
> similar to that as Monongah. I made a thorough
> investigation of the French mine, and my knowledge
> gained there will be of help to me here.

Cafinel's testimony encouraged the general belief that the
explosion occurred first in Number Eight, spreading with

intense heat and speed to all the other parts of the two mines.

A group of inspectors from Ohio corroborated Cafinel's opinion about the initial point of explosion. The Ohioans adamantly refused to believe in the theory of the runaway trip: "We made a close inspection of this wrecked trip and its surroundings, and failed to find any evidence of fire, or of any justification for the belief that the explosion originated at that point or by contact with the electric wire."

A similar group of mine inspectors came from Pennsylvania to aid in gathering evidence for the coroner's jury. Its opinion did not differ materially from most others: the explosion occcurred in Eight because of a blown out shot, and then spread through the mines by the ignition of coal dust. The runaway trip probably intensified the explosion by raising up additional amounts of coal dust. The explosion and the runaway trip, however, were coincidental.

It was only when the West Virginia inspectors filed their report that a cleavage of opinion developed. Two of the dozen inspectors favored the runaway trip theory. They argued that bituminous coal dust, by itself, had a very explosive nature and did not—as some argued—have to come into contact with gas, methane or otherwise, to ignite. Given the speed of the loaded trip when it reached the bottom of Number Six slope, it was perfectly logical to surmise that tons of dust would instantly be lifted throughout the mines—and that its volatile condition would produce disaster.

The question of the runaway trip *versus* the blown out shot got a huge amount of attention in all the newspapers. And, as it turned out, Coroner Amos's jury spent many hours debating the two theories, and hearing testimony about them.

Dr. Henry M. Payne, Professor of Mining Engineering at West Virginia University "proved mathematically" that the runaway trip was not the culprit. The runaway cars, he said, were doing thirty miles an hour when they reached the opening, going backwards. Halfway down the slope, at his reckoning, they had reached fifty MPH. It required, asserted Professor Payne, twenty-five seconds for the runaway trip at those speeds to reach the bottom of Number Six slope. The explosion, however, came seventeen seconds (according to witnesses with whom Payne spoke) after the runaway trip had re-entered the mine. Therefore, it was hardly half-way down when the explosion occurred.

Attorney for the Fairmont Coal Company, George M. Alexander, offered to reproduce the runaway trip as proof that it could not have caused the accident. His company, he said, was quite willing to turn loose another fully loaded nineteen cars and let them pile up at the bottom of Number Six slope. The jury did not take up Alexander's offer, either convinced by his argument, or non-plussed by his audacity.

The jury's verdict was to the point, and imprecise: the explosion was "caused by either what is known as a blown out shot or by the ignition...of powder in mine Number Eight." It recommended a number of things: the creation of a federal Bureau of Mines, the use of safety or flameless powder, the use of clay in tamping shots, and the hiring of additional mine inspectors.

Hence, the coroner's jury gave no credence whatever to the theory of the runaway trip. And there the matter rested for the next seventy years. All through that period, however, there was one person who stoutly refused to accept the coroner's verdict, but until only recently (1978) could get very few people to believe him.

Lester Trader, of McKeesport, Pennsylvania, was twenty-two-years old in 1907, and was a fire-boss at Number Six. "The fire-boss," he said, "was [also] the sprinkler, and we wet down the dust where we could, which wasn't much. Of all the damn crazy ideas. How much of a mine do you think one man with a horse and water cart can sprinkle in one night...? He also had the duty of placing warning signs at sites he found to have accumulated unsafe quantities of gas.

After the explosion, Trader testified before the coroner's jury, but he acted upon the orders of Fairmont President Watson: "Tell the truth, but volunteer no information." This admonition, given not only to Trader, but to all other Monongah witnesses as well, effectively scotched much of the information that could have come before the jury, but did not. After all, there were still jobs to protect.

Runaway trips, according to Trader, were fairly commonplace occurrences. Just three weeks before the big explosion, for example, a pin broke and sent four empty coal cars slamming back into the slope of Number Six. On this, as on previous occasions, the grinding wheels sent out sparks which ignited the suddenly agitated coal dust. A ball of fire was created which rolled down the heading, away from the wreck.

These balls of fire caused by the wrecks of runaway trips, according to Trader, always burned out after rolling a few hundred feet. Consequently, not too many miners overly concerned themselves about any explosions that might occur from runaway trips.

All right then. What made the December 6 occurrence different from all the previous ones? Here is a letter written by Trader in June, 1972, to the Catholic Churches of Monongah:

> An additional piece of evidence not presented at the inquest was the presence of a [fifty pound] box of dynamite, a box of caps, and a roll of fuses, in the maintenance locker room...about 1,000 feet down the main heading from the scene of the wreck. The dynamite had been taken into the mine on the Saturday prior to the explosion for the purpose of blasting [away some slate on one of the faces]. But when the crew arrived at the...face..., someone had neglected to place empty cars for loading the slate, and the work had to be postponed until the next Saturday night. [The explosion occurred the following Friday.] I was able to check this location later and found evidence of a terrific force—strong enough to blast a hole twelve inches deep in the slate...and tear the wood tops completely off two water cars that were parked in the rear of the crosscut.

Trader went on in this letter to say that a team of German mine experts noted the dynamite explosion, and surmised that this was the cause of the tragedy, and said as much to the jury, only to be ignored.

Trader's account of the explosion exonerated the miners in Number Eight, accused of neglect and carelessness in the matter of blown-out shots. On the contrary, sloppy supervision caused the accident. It says little, indeed, for a management who would allow a large crate of dynamite, with several sticks capped and fused, to stay in a conspicuous yet vulnerable place for an entire week. To have removed the dynamite before the blasting that was scheduled for Saturday, the 7th, would have involved a special work crew, losing time in actually producing coal. Even so, one official told the jury that the company never sacrificed safety in the interest of economy.... Incidentally, the coroner's jury thoroughly cleared the company of any negligence that might have contributed to the disaster.

(As stated earlier, Trader spent many years trying to draw attention to the dynamite theory—though this was no "theory" for him—it was fact. In a letter, May 16, 1973, to Arnold Miller, President of the UMWA, Trader said: "Some time ago, I tried to interest the parties in this matter in clearing up the missing links and get the real facts on the official records. I wrote to the Marion County [West Virginia] coroner, and then to the West Virginia Bureau of Mines, and later to the Consolidation Coal Company. Not one of these parties even took the trouble to acknowledge receipt of my letters...." Trader died in July, 1978, in his mid nineties).

Trader's explanation of what happened at Monongah is certainly as credible—maybe more so—as the coroner's jury's. It is true that Coroner Amos and his jurors took the testimony of several workers at the Monongah mines, including Trader's. He told them about the runaway trip three weeks before the big explosion, but apparently omitted the details about the box of dynamite, saying instead that not enough dust was suspended at that time to cause an explosion. It is also true, however, that the great bulk of evidence—the evidence upon which the official findings and verdict were placed—came from "experts" who had never even been near the Monongah workings. The findings of high ranking inspectors from European countries (with, apparently, the exception of Germany), Ohio, Pennsylvania, and West Virginia, were given greater weight than several of the men whose daily lot it was to trudge in and out of the darkness, grubbing a living for themselves and their families.

The coroner's clearing of the company of any liabilities may have put a legal cap on the Monongah horror, but it definitely did not stop the surmising. For months afterward newspapers and magazines continued to flood their readers with explanations and opinions, many of which demonstrated downright hostility toward the coroner's jury.

The explosion provided plenty of ammunition for the muckraking press in the United States. Showing West Virginia's poor reputation in safety, Paul Kellogg, writing in *The Charities and the Commons* (the magazine issued by the organization of the same name), offered these statistics: in 1904, five inspectors had to cover 629 mines in West Virginia, making 1,446 visits altogether. This meant that "if an inspector should merely go to the main entrance of each mine in his district four times a year, it would consume every day...."

Moreover, inspectors only had the authority to recommend, not enforce. Coupling the weaknesses of the official inspection system with the company's profit motives produced perilous work indeed in the West Virginia coal fields during the first years of the twentieth century. It must be pointed out, also, that the individual miner of that day was paid in accordance with how much coal he produced—a practice that definitely encouraged recklessness. Sometimes a lone miner was willing to jeopardize the whole operation just because he wanted to earn a few extra dollars.

Not enough coal mine training and study had been developed, said Kellogg, in West Virginia. New equipment (particularly electrical) was quickly turned over to unskilled, green laborers. The geologists, both state and federal, had performed lackadaisical studies of the mines' interiors, being more interested in the "pterodactyls and other friendly beasts that sported in them than of the life and death risks of the men who work there...." This geological nonchalance, said Floyd W. Parsons, in *The Engineering and Mining Journal,* produced a distinct false sense of security—the belief that any explosion would be confined to only a few areas of a mine, and not the entire workings.

The explosion caused some magazine writers to report an increase in prejudice against foreign miners. One such was Edgar Allan Forbes, in *The World's Work*. He pointed to the growing belief that the aliens caused most of the problems: "What is the matter with the foreign miner?' I asked an intelligent young fellow resting in a coal car. 'They can't stand it,' he said. 'They can handle a pick all right, but when anything happens, they lose their heads.' " Forbes, thus gave the opinion that "unless something can be done to safeguard the lives of the men in the mines, the native American miner will shortly disappear from the hills of coal.... With a steady stream of muscular aliens flowing into the country, the operators have not concerned themselves much about the natives. If the mines go deeper and the call for intelligent labor becomes stronger, [however], the American will be needed."

A final theme inspired by the Monongah explosion was that the disaster should not be considered in a vacuum. It was not a thing by itself, but by extension showed involvement in the problems of economics, immigration, social programs, and industrial education. Certainly, Monongah pointed out that industry can and does develop faster than any social mechanism which is called upon to cope with its disasters. The MRC was a partial antidote to this condition but, as we have noted, it became increasingly obvious that the federal government would have to take a direct hand in such calamities.

As stated earlier, the coroners' verdicts for the first two blasts of December, 1907, were imprecise. This trend of imprecision continued for the third one, Yolande, and by the time of Jacob's Creek, it was definitely entrenched.

A team of company experts at the Darr mine blamed worker violations, while the inspectors of the State Department of Mining cited past instances of supervisory neglect—some going back as far as the Spring of 1906.

For example, on May 22, 1906, John F. Bell, Inspector for the Thirteenth District, found unsafe accumulations of gas at Darr, open lights, indiscriminate shot firing, and slate piles that obstructed entries and tracks. He stated on this, and subsequent occasions, that the situation at Darr was extremely dangerous.

Sometimes at Darr, things got completely beyond the control of supervisors. In March, 1907, August Nemeck and B.R. Peters, both lawmen in the area, came to Darr to find a man they wanted to arrest. According to reports, Nemeck and Peters wandered through the various passageways, using naked lamps. Darr Superintendent Archibald L. Black explained in a letter to Bell: "They [the two policemen] have been in Section 27...with naked lights without the permission of any of the mine officials. When they were informed...that it was a violation of the mining laws [to come into the mine without permission]...they began to abuse the mine officials, and informed them that they would come into this or any other mine they chose. They also threatened to arrest the assistant mine foreman, for claiming that he thought they had done wrong. Their authority is such that our positions...are useless in regard to the safety of our men." What with reckless workmen, indulgent supervisors, and arrogant constables, it is a wonder that Darr did not explode months before it did.

Coroner Wynn's jury was torn so that it could only say that the disaster occurred "as a result of explosive gas ignited by an open lamp, at a point which we are unable to locate owing to a difference of opinion between inspectors and experts." It did point out what was well known to all that "the explosion was augmented by coal dust which extended throughout the mine." Despite its uncertainty of the exact cause or the initial location, the jury found "that the bituminous laws of Pennsylvania were fully obeyed by all those having charge of the Darr mine...." It closed its report by recommending better ventilation systems for Darr, greater care in shot firing, and a Bureau of Mines on the federal level.

All of the coroners' juries accepted the casualty rate as reported by company and government officials. For Naomi it was thirty-four; Monongah, 361; Yolande, fifty-seven; Jacob's Creek, 239; and Carthage, eleven. No credence whatever was given to the general belief and, in some cases, public testimony, that these figures were drastically low. The belief persists to this day among certain segments of these mining communities that more than 702 men died in American coal mines in December, 1907.

A chart (somewhat facetious) of the work of the various coroners' juries arising from the dreadful events of 1907 might look this way:

	Did an explosion occur?	Why, Exactly?	Were any miners at fault?	Was Company at fault?	What can be done?
Naomi	Yes	Unknown	Yes	No	Bureau of Mines
Monongah	Yes	Unknown	Probably	No	Bureau of Mines
Yolande	Yes	Unknown	Probably	No	Bureau of Mines
Jacob's Creek	Yes	Unknown	Probably	No	Bureau of Mines
Carthage	Yes	Unknown	Yes	No	Bureau of Mines

It is ironic, to say the least, that while the juries in every instance exonerated the company, they also counselled that the U.S. government step into the picture, because the states had proven inadequate to the task of mine safety. When the Bureau of Mines did begin to operate (and after it received enough power to enforce its recommendations—a procedure that took several years), it most assuredly regulated and disciplined more companies than individuals.

The explosions caused a flurry of activities from state and federal governments, reflecting Edgar Forbes' opinion that "not until grave economic conditions begin to imperil the store of coal reserves or threaten a reduction of dividends do legislators and stockholders seriously contemplate sweeping measures for reform." In Pennsylvania, an Accident and Pension Fund was discussed among the state legislators. Initiated by Samuel A. Taylor, an engineer and coal operator, the fund would care for survivors of miners killed in accidents, and support those who were crippled. Coupled with the suggested fund, however, was a demand that the State eliminate its Capital Stock tax, and settle for a tonnage tax on actual production. Some owners believed that companies which had large productions were not taxed in proper proportion to the production of smaller operators. A tonnage tax would have an equalizing effect. Nevertheless, when it was generally learned that the Taylor Plan had these tax features, the feeling arose that it was simply another effort by the companies to trade safety and humanity for a bit of extra profit. This may not have been their motive at all, but it was the image they projected. At least it can be said that the Pennsylvanians suggested *something*, while Alabama and West Virginia did very little in the months immediately following the explosions to lessen the suffering.

The state and federal solons were advised by numerous technical writers. One such was J.L. Dickson, who, writing for the *Dispatch,* wanted laws for mandatory dampening of mines, for hauling out dust and dirt accumulations, and for prohibiting electrical machinery in gaseous areas. Another, C.E. Mayers, in an editorial for the *Wheeling Intelligencer,* suggested uniform standards for mine ventilation. He believed there was too much air at Monongah, especially Number Eight which, regardless of the amount of dampening, kept the dust too dry, and thus too susceptible to an explosion. An essay by Clarence Hall and Walter Snelling, *Coal Mine Accidents: Their Causes and Prevention,* got a wide currency. These authors, like several others, emphasized the explosive qualities of coal dust, a factor that had received little attention until only recently.

Some sources, however, disdained legislative activity in

the area of coal-mine regulation. Typical was *The Parkersburg* (W.Va) *State Journal,* which editorialized that "a legislature is about as well fitted to cope with a mine as a muskrat with a stone dam." The Fairmont *Times* joined in with the observation that legislative office in West Virginia carried more disgrace with it than honor. Both miners and owners—to say nothing of the general public—had been maltreated by *status quo* politicians, who were responsible for the absence of safety regulations. Another paper expressed displeasure at the growing tendency for companies and states to turn to the federal government in times of industrial crisis. Certainly, the editor pointed out, New York Governor Charles Evans Hughes would not ask Washington to help if, say, large numbers of New York banks were to default....

Undaunted by these, and other unfriendly reactions, the politicians forged ahead, as it became increasingly fashionable to take up coal-mine reform. In Washington, for example, Representative John G. McHenry of Pennsylvania, introduced a bill to levy a tax of one cent a ton on all coal mined in the country to create a fund for persons injured in mines and to provide for widows and orphans left from accidents. Representative George C. Sturgis of West Virginia, wanted to appropriate five thousand dollars to each of the land-grant colleges in the United States, to establish Schools of Mining. This amount was to be increased by five thousand dollars each year until it leveled off at twenty-five thousand. The Congressman was chided for thinking "the loss of life can be prevented by having everything pertaining to mining taught in the colleges."

Congressman George F. Huff, of Pennsylvania, a mine operator, was appointed as the Chairman of the House Committee on Mines and Mining. He said, "I don't think mining in this country is under such radically different conditions from those in Europe to cause the greater loss of life.... If they have better laws in Europe, and better practices, we ought to know the fact, and I'm going to do what I can to find out. Even from the cold, practical business point of view, it is cheaper to avoid accidents...."

Edward H. Madison, a Congressman from Kansas, wanted to establish a broadly based "Commission to Investigate Recent Mining Disasters." He pointed out that the public apparently thought railroading was the most dangerous occupation in the United States, and that this misconception caused the foot-dragging in mine safety. Bureau of Labor statistics, however, said Madison, indicated that of each one thousand railroad workers, two and a half were killed annually; of each thousand miners, three and a half men lost their lives. It was Congress's duty, he believed, to educate the general public on the real conditions to be found in America's coal mines.

In support of his commission idea, Madison said: "It is greatly to the discredit of this country that while the sacrifice of human lives in our mines has been increasing, it has been decreasing in some of the countries of Europe, notably France and Belgium—in both of which coal mining is a leading industry. A commission authorized by Congress to investigate the causes of the recent disasters could determine such causes and...ascertain some means toward tending to eliminate them." No such commission was ever established; the dialogue, however, about coal mine disasters never got to a higher point of intensity than in January and February, 1908.

(But, even as these people spoke in Washington, a Special Committee in Pennsylvania appointed by State Mining Chief James E. Roderick, reported several mines as allowing unsafe shooting practices, and not having proper ventilation. One of these was Naomi.... Another was Darr.... And the reports kept coming in throughout 1908 and on into 1909. The dreadful experiences of December, 1907, had apparently made little impression on certain miners and operators).

The Congressmen's voices gave added urgency to requests and demands that had been coming for some time from the Executive Branch. The Interior Department asked Congress for $60,000 to investigate ways of preventing mine accidents. It suggested, too, that a "test station," of at least two acres of land, be provided to serve as a laboratory to produce "controlled disasters" so that the real thing could hopefully be avoided.

The Department of the Interior was especially persuasive in its requests and recommendations about powder. Joining dozens of newspaper editorials around the country, the Department condemned the unbridled use of black powder, and also the government's failure to regulate it. George Otis Smith, speaking for the Department, pointed out that the manufacture, sale, and use of explosives in foreign countries were closely regulated, and they had better safety records to show for it. Obviously, a federal law was needed in this country to harness the wanton use of dangerous powders.

Interior noted another problem, however, in connection with explosive powders. Because of the December tragedies, some mining companies had begun to buy "safety" or "flameless" powder. It was many times a case of "caveat emptor," as the Department explained: "...An explosive known and advertised as a flameless blasting powder has been found on analysis...to be simply common black powder, in which ordinary bituminous coal has been substituted for the charcoal ordinarily employed. This explosive is not only not flameless, but it is undoubtedly a more dangerous material than ordinary black powder, and the use of the term 'flameless' in such a case is misleading...." A sample of "safety" powder was taken from the Whipple, West Virginia mine. It was a dynamite of fifty percent nitro-glycerine content, "a grade of material most unsuited for coal mine operations,...the use of which would be forbidden in any other country."

President Roosevelt had long railed out at "malefactors of great wealth," who, in the name of "free enterprise," sent hundreds of men to their graves each year, and crippled thousands more, all for monetary profit. He was scornful, too, of a court system that slapped miners and other laborers with injunctions the very instant they raised their voices against unfair and unsafe working conditions. To be effective, he said, an injunction must be summary; thus it was always administered without benefit of a jury—a practice that undermined the democratic process.

The President was adamant in his recommendations for a Bureau of Mines. Long before the horrors of December, 1907, he had urged Congress to create this regulatory agency. Congress had always hesitated, however, fearful that its constituency would yell "socialism." Roosevelt emphasized that the Bureau would not be so much "new" as "re-organized," noting that several agencies in Interior and Commerce presently carried out mining studies and activities. The Bureau, he felt, could become a clearing-house, where the results of experiments could be distributed to all the coal-mining states. Thus, a co-operative effort would be created between federal and state governments, with the almost certain results of saving lives.

Roosevelt gave an impassioned appeal for a Bureau of

Mines on January 22, 1909. This certainly put Congress into a "speedy" frame of mind, for it took almost sixteen more months before it gave—on May 16, 1910—the final touches to the long awaited bill. Congressman William Wilson of Pennsylvania, later to be the first Secretary of Labor—under President Wilson—was its chief sponsor.

The Bureau, whose director was to receive an annual salary of $6,000, was to "investigate, analyse, and test coals, lignites and other mineral fuel substances." It was expressly charged with conducting experiments to reduce the number and severity of explosions and other accidents in coal mines. It carried out this function at test facilities built in Bruceton, Pennsylvania. (Today, of course, it has dozens of research stations scattered throughout the United States.)

The first director of the Bureau of Mines was Dr. Joseph A. Holmes, a geologist who had long worked for the Interior Department. He earned the enmity of the UMW, however, for being too "production oriented," for hiring political favorites, and for knowing too little about coal mining. One disgruntled coal miner, Charles Tinlin, told the *UMW Journal* that the director was a "big joke."

On the surface at least, the Bureau of Mines certainly did not prevent accidents during the first several years of its existence, as these chilling figures indicate: from 1870 to 1910—while there was much demand for a Bureau, 42,446, or a few over a thousand a year, died in coal mine accidents. During the next forty years, however, 71,102 men, approximately 1,800 a year, perished in America's coal mines. It was only during the 1950s that fatality rates dropped significantly. From 1952 to 1977, there were 7,361 deaths, an average of slightly over two hundred a year.

One possible explanation for the continuance of tragedy is that the actual number of coal mines was greater from 1911 to 1951 than during 1870-1910. Thus, the real percentage of fatalities for the two periods was not as discordant as it might appear at first glance. From the fifties on down to the present, natural gas was the chief supplier of heat and energy, causing coal mine production to fall on evil days; it has been only since the fuel crisis of recent times that coal is again looked upon as our salvation.

Another possible reason why the fatalities continued after the Bureau was established, was that the Agency, right from the beginning, did not concern itself solely with trying to prevent coal mine accidents. A major part of its activities, for example, was taken up in experiments of various kinds. It researched the separation of helium from natural gas, it created synthetic fuels from coal and oil shale, and made possible the commercial production of metals like titanium, zirconium, and hafnium. In latter day times, the Bureau has been instrumental in air pollution research, and finding ways to ameliorate the ravages of "black lung." To be sure, all these objectives are highly laudable, but more than one miner regarded them as peripheral—they did not aid materially, at least for the first several years—in stopping the awful, sudden, snuffing out of human lives in the dark recesses of the earth.

It must be noted further that for several years the Bureau really did not have sufficient authority to enforce its findings. Only in contemporary times (1952) have companies and state governments "sat up and took notice" when the Bureau speaks. Regrettably, it took many mine accidents and loss of life before Congress got around to endowing the Bureau with enforcement authority.

It is both a blessing and curse that the human race has such short memories. The Pittsburgh *Dispatch* said: "While the memory of the previous disasters is fresh, all are watchful and alert against the causes that [could] produce [another]." It went on to congratulate the owners, superintendents, and miners for their new-found efficiency in ventilation systems, and for shutting off gaseous parts of mines. The paper closed its article, however, by asserting that "the memory becomes cold. The absence of new disasters engenders a false security. One or another [of the mine personnel] relaxes the constant watchfulness until finally new disasters come to enforce the lesson. A never sleeping care is the price of safety in mining, and even then does not always secure it."

At Monongah, the *Dispatch's* implications were set in motion during the closing days of January, 1908. The Fairmont Company had advertised for replacements, and this turned Monongah into a boom town. Superintendent A.J. Ruckman received so many applications that he could not keep up with them. They came from Americans, English, Irish, Scots, Italians, Slavs, Poles, Greeks, Spaniards, French, Lithuanians, and Swedes. The new workers were promised that every precaution would be taken to prevent another explosion. But these words were familiar—they had been said before. Flameless powder would be used. The dust would be kept damp. Accumulations of waste would be taken from the mines at frequent intervals. This was all just fine, for here were men who needed jobs—there were many more men, actually, than there were jobs.

Number Eight started running around January 25, and Number Six was operative by February 1. Thus was everything returned to normal, and once more all was well.

But was it? Certainly not for a man named George Riggins. His brother, David, was lost on that dreadful day, December 6, and George kept dreaming that David was reposing in one particular spot in Number Eight. He could not rest until he and a brother-in-law went to the place. Sure enough, there was the huddled form of David, in a rather advanced state of decomposition. He had lain there, undetected, for six weeks.

Not so well, either, was it for a group of miners in Six who, one day at work, came upon a human heart left by the explosion. It was in a crevice in the rear of an abandoned working. The explosion blew out the timber holding up the crevice, and replaced it with the heart, and only the heart, of a worker. And for the next several months, chilling reminders that indeed not all was well kept confronting the new miners at Monongah.

But the miner's heart was not really the only one that was found at Monongah. Because of the painful suffering of widows, the cruel deprivations of children, and the utter hopelessness of surviving parents, America found its heart, too, and expressed an interest in dictating that never again would there be a dreadful month in the United States.

Sources

Coroner's Report for Marion County (Monongah) West Virginia, January 15, 1908.

Department of the Interior, *The Bureau of Mines: Its Mission and Programs.* Undated Pamphlet

Fairmont *Times,* December 24, 1907; January 8, 9, 23, 28, 1908.

Forbes, Edgar Allan, Article on Monongah in *The World's Work.* Vol. 15 February, 1908, pp. 9929-31.

Haas, Frank, *The Explosion at Monongah Mines,* Bulletin No. 11, Fairmont Coal Company, 1910.

Hall, Clarence and Walter O. Snelling, *Coal Mine Accidents: Their Causes and Prevention.* Washington: Government Printing Office, 1907.

House Document 523; Letter from the Department of Interior, Geological
   Survey, January 17, 1908.
House Report 678, February 5, 1908. "Commission to Investigate Recent
   Mining Disasters." 60th Cong. 1 Sess.
Interview with Lester E. Trader, May 25, 1978.
Kellogg, Paul U., "Aftermath at Monongah," in *Charities and the Commons,*
   January, 1908, pp. 1313-28.
MSS. Darr, Pennsylvania State Department of Mines.
Osburn, Elliott, *Mission Technology: The Story of the Bureau of Mines,*
   New York: Cities Service Company, June, 1972.
Parsons, Floyd W., "Disaster at Monongah Coal Mines No. 6 and 8," in *The
   Engineering and Mining Journal.* December 14, 1907, pp. 1121-23.
Pittsburgh *Dispatch,* December 7, 24, 31, 1907; January 17, 21, 27, 1908.
Report, U.S. Geological Survey, Department of the Interior, January 11, 1908.
Roosevelt, Theodore. Message to Congress, January 31, 1908, in Richardson,
   Messages of the Presidents. Washington: Bureau of National Literature and
   Art, pp. 7128, 7267.
Stafford, Sam, "America's Worst Mine Disaster," in *Mine Safety and Health,*
   February-March, 1978.
Swenarton, R.O., Letter to the author, March 29, 1978.
Trader, Lester E., *Letter to Catholic Churches of Monongah,* June 11, 1972;
   to *Governor Arch Moore of West Virginia,* September 16, 1974; to *Editor of
   Fairmont Times,* October 6, 1971; to *Arnold Miller,* May 16, 1973.
United Mine Worker's Journal, No. 3; January 16-31, 1974.
Veneman, Wayne E., Letter to the author, undated, 1978.

Perspectives III

Sixties—Now

Ironically it seems, most major mine accidents occurred just before festive holidays. December is the worst month for explosions, with November taking a heavy toll as well. There have been so many "black Christmases" that it is a melancholy task even to describe, let alone experience them. Likewise, dozens of Thanksgiving celebrations have been erased because of events in deep, underground caverns.

Such was the case near Farmington, West Virginia where, on November 21, 1968, a series of explosions took the lives of 78 men. It had been seventeen years since the tragedy at West Frankfort. Partly because of the time lapse, and careless memory, the Farmington explosions caught everyone off guard, and the event soon reached major, world-wide proportions.

It happened at 5:40 a.m., just when the midnight shift had two hours to go to quitting time. It was Mine Number Nine, and it belonged to Mountaineer Coal Company, a subsidiary of Consolidation Coal Corporation, now based in Pittsburgh. It was eight miles from Monongah.

Throughout the day, "pop-offs" (comparatively small explosions) recurred, sending billows of dirty white smoke from all the portals, indicating that fires were still sweeping through the tunnels. Under these conditions, rescue attempts were impossible.

Champion's Store became the gathering point for the thousands of people who swarmed into the area. Newsmen, television crews (this was the first time TV was used to such an extent to cover a coal-mine disaster), state and federal mine officials, women and children crowded into the little structure, waiting for some news—hoping it would be good.

Reporter Niles Lee Jackson filed a story for Associated Press about Mrs. Barbara Toler. Her husband, Dennis, was in the mine, and she would not leave Champion's Store, though

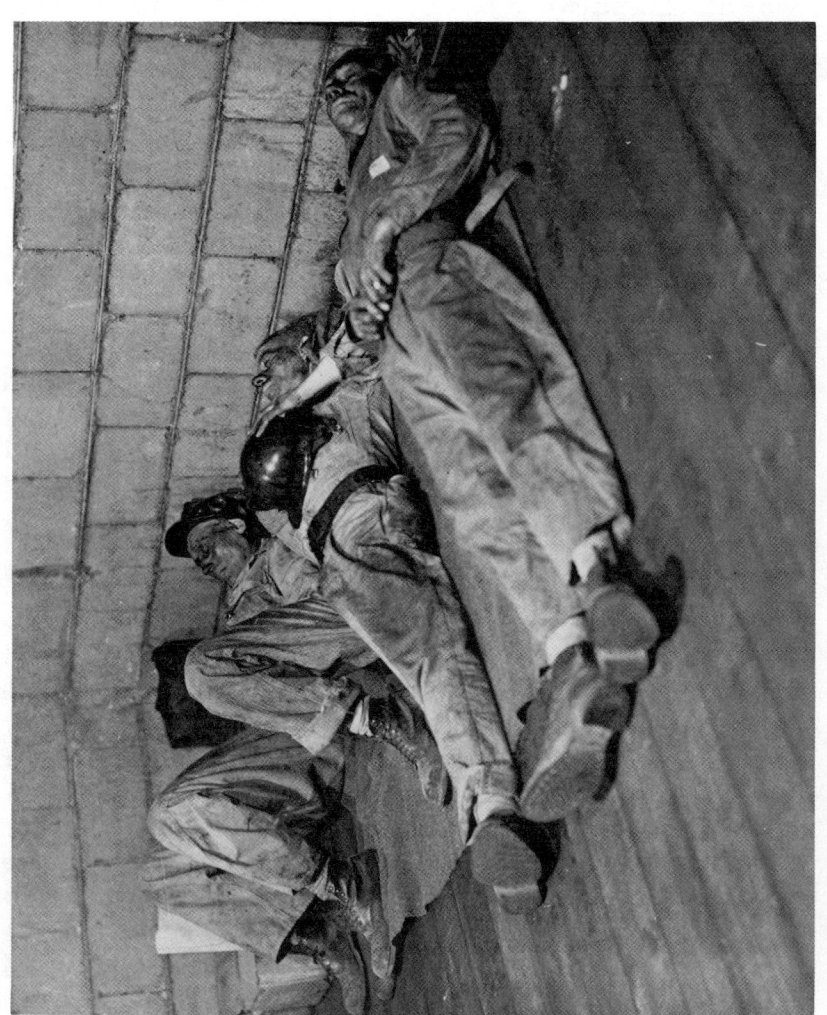

Exhausted rescuers take a break after an un-named explosion, ca. 1940s. Photo courtesy, UMWA.

she was just on the verge of delivering a baby. She would not go to the hospital, because "they'll keep me four or five days." If her baby were born on the mine grounds, she said, it would be declared a "dirty baby," and thus not allowed in the hospital. (Later, Red Cross representatives convinced her to have her baby at a hospital.)

Barbara gave a touching picture of what it was like to be a coal-miner. "We owe the company store and other people so much Dennis has got only two paychecks in two years." His pay had been garnisheed. He took cabbage sandwiches with him into the pits each day, gave up smoking as an economy measure, and doubled as a grave-digger in nearby cemeteries. He averaged two hours of sleep a day. Barbara said, "That song, '16 Tons,' sure is true for him. That's sure his song if any is."

There were other stories: of Hartzell Mayle leaving his wife, Juanita, and their sixteen children; of dry-eyed wives collecting their lost husbands' paychecks, while the roars of explosion filled the background, and TV cameras rolled; and how, when bad luck starts, it just keeps building onto itself; a week ago a department store in Farmington had burned, killing four people.

There were some survivors, and they had stories to tell. Most were ill and vomiting as they ran from the mine, and two were bleeding from the eyes, apparently as a result of concussions. Lewis Lake said, "The Lord was with us. He brought us through, because we couldn't see a thing." Nezer Vandergift reported that just after the explosion, he had never "seen dust that bad. It was thicker than when dynamiting." Alva Davis spoke of his friend, Paul Anderson, who apparently ran the wrong way after the first explosion. "I left a note in case he (Paul) comes back, telling him where we went. I wrote it on the coal-car with my finger in coal-dust." Alva went on to say that "men underground are like brothers. We try to take care of each other. I'm afraid for Paul."

There were plenty of officials on hand to give statements. West Virginia Governor Hulett C. Smith, said, "We must remember that this [coal-mining] is a hazardous business and what has occurred here is one of the hazards of being a miner." UMW President W.A. Boyle said that Consol was "one of the best companies as far as cooperation and safety are concerned." Then he added: "It looks very dark to me [about

Rescue crew arrives at Benwood, W. Va. April 28, 1924.

rescuing the trapped miners]. But I will be the last person in this room to give up hope."

Interior Secretary Stewart L. Udall said he would convene a conference in Washington in ten days to study ways of preventing any further "mishaps" in the nation's coal mines. Claiming that everyone concerned, including his office, had become too complacent, Udall declared that "such mine accidents [as the one at Farmington] aren't acceptable in a country with the technological know-how and wealth like the United States."

In a speech at the University of Notre Dame, consumer advocate Ralph Nader sounded off: "The situation of the coal miners in this country is reminiscent of medieval times. The political representatives in Washington, D.C., and in states like West Virginia and Kentucky represent coal; they do not represent the coal miners.

And in Washington, West Virginia Representative Ken Hechler indicated his intentions of resurrecting a bill recommended by President Lyndon B. Johnson that had been denied by the 90th Congress. It would enable the Secretary of the Interior to develop and issue new mine safety standards. And so it went.

By the 24th, on the fourth day after the explosion, one

rescue unit reached as far as 4,500 feet into the mine. They could see signs of a concussion, and also they detected large amounts of explosive gases, and vast quantities of floating coal dust. Because there had been at least a dozen explosions prior to the rescuers' entry, and because another was imminent, they were ordered out of the mine.

The next effort at getting to any possible survivors was to drill some holes into the mountainside, straight down into the shafts. Three drills, each measuring three inches across and equipped with diamond-head bits and tiny, sensitive microphones "that could pick up a whisper at one hundred feet," were plunged into the depths. The only sounds a waiting world heard were the gurgle of water and pebbles dropping against a microphone.

A somewhat heated exchange occurred between William Poundstone, Consul's Vice-President, and Tony Megna, a high school principal from Columbus, Ohio, whose brother was in the mine. Reporter Jackson and his partner Joe Kroviskey, filed this "conversation:"

> "Why don't you get more drills in here?" asked Tony Megna. . . . Megna suggested that rescuers drill over each spot "where they might think there's a section of men barricaded in.
>
> Poundstone replied, "Remember, we've got an area of enormous proportions. We've just tried to spot these."
>
> "Yes," Megna said, "but we've got hundreds of drills in this state. Why not bring them in? It wouldn't take that much money."
>
> Poundstone shot back, "It's not a matter of money, I can assure you."

That nerves were becoming frayed and patiences wearing thin was demonstrated further by discussions of whether or not the mine had followed recent safety recommendations. The cause of the explosion was fairly well conjectured as early as the second day: it was probably a mixture of methane gas and coal-dust. Well, nothing new here. Such mixtures had been killing coal miners for a long time. The big question was: who was responsible? And in trying to find out, nothing new, either, was unfolded.

Company officials reported that 125 tests had been taken at Farmington in recent months, and that only three had shown coal-dust above the maximum "safe" level of sixty-five percent. The mine had been found in good shape by all inspectors, federal and state, in 1967. Its last inspection was in August, 1968, and a few questions arose as a result. These had to do with controlling coal dust, and properly covering electrical equipment to prevent arcs that might ignite the coal dust.

M.W. McManus, Bureau of Mines inspector, ordered after the August inspection that excessive accumulations of "fine coal" and spilled oil be removed. State inspectors Walter Miller and John Ashcraft gave the directive, and ordered that additional non-volatile rock dust be spread.

All of the officials later insisted that these orders were obeyed. William Parks of the Bureau of Mines maintained that if the mine had been unsafe, the Bureau would have closed it. Lewis B. Evans, an official of the UMW, said reports "indicate very conclusively to me that the mine, at least when the federal inspectors left, was in a safe condition. Whether it was safe in the months of September or October,... I can't say." It was obviously not safe during the month of November. No distinct answer has ever been given on who was responsible. If tragedies past are any guideline, and surely they are, there is plenty of blame to spread around, and not one soul willing to take it.

As time wore on, and as the explosions continued (by the 27th, there had been fourteen of them) the fearful thing that was on everyone's mind was articulated: would the mine have to be "capped," that is, sealed? That would certainly stop the fires (after tons of fist sized chunks of limestone being poured in had failed), because it would deprive them of oxygen. It would also spell doom for any of the 78 miners who might still be surviving (eighteen days is the record for staying alive in a coal-mine).

Of course, the officials kept saying they would not do it. On the 22nd, Poundstone said there were no plans to block up the mine, although at that very minute workers were stockpiling cinders, block, and mortar on the grounds. Two days later, a Consol' engineer, Alder Spotte, said that sealing the mine "is not in our minds." At the same time, however, James McCartney, Public Relations Director for Consolidation, said there would be "no further exploration of the underground areas...pending further information." For the next five days, drilling continued through the mountain, huge amounts of limestone balls were thrown down the shafts trying to extinguish the blazes, and frequently an official would tell the media that no plans existed for sealing off the mine—although everyone could see that the stockpiling of materials continued apace.

Just at daybreak on Saturday morning, November 30, huge trucks rumbled into the Farmington mine grounds. They carried cement, timbers, limestone, and mortar. Any relative in the area was quickly taken away from the scene, as workmen began the gruesome task of sealing up the mine. Thus was the great fire stopped, and with it, any chance whatever for the entombed men. The officials had met late on the night of the 29th, and concluded that any survival was an absolute impossibility. Thus, they gave orders to block up the mine.

The relatives met this decision with a stoicism possessed perhaps only by coal mining families. Many of them now wanted to know when the mine would be un-sealed, and proper burials provided for their lost husbands, fathers, and sons. The coal company made an agreement with the widows that the mine would be un-sealed when the fires were completely out, and when it was "safe, feasible, and practical," for recovery operations to begin. On October 23, 1969, the mine was un-sealed, and the first body brought out. Additional bodies were recovered until December 2, 1977, when it was decided that the workings were deteriorating so rapidly that a real risk of further tragedy was possible. Retrieval of the lost men, therefore, came to a complete halt. On April 19, 1978, the mine was permanently sealed; nineteen miners' bodies were still in it.

The Farmington explosions caused Congress to pass, a year later, the Federal Coal Mine Health and Safety Act. It increased the power of federal inspectors in closing down mines "in cases of imminent danger," and it provided for a federal dust control program. The UMW, however, lamented the failure to enforce this law, as its *Journal* bitterly complained:

> Unfortunately, Richard Nixon and the political hacks he appointed to administer the law showed little interest in making the coal operators obey its standards. In the six years after the Act went into

effect, 1,000 miners were killed on the job. No one ever spent a day in jail or paid a penny in criminal fines for violations connected with these deaths. The government routinely settled for small fractions of the civil fines assessed for millions of safety violations.

In recent times, there have been instances of coal miners' causes being taken up by inspired lawyers, and this has made both operators and government more wary about mine accidents than they once were. A case in point is the Scotia Mine in Kentucky, where two explosions in March, 1976, took 26 lives. The Scotia Company was owned by the Blue Diamond Coal Corporation of Knoxville, Tennessee. According to its attorney, Bert T. Combs, Blue Diamond gave each widow from the first blast $5,000, "started paying maximum workmen's compensation, paid medical expenses for two or three of the women and considered setting up a foundation to maintain and educate the children." This showed, said attorney Combs, that the company had the welfare of the survivors at heart.

A Washington based attorney, however, disagreed with Combs' stance. Gerald Stern said Blue Diamond was "indifferent" in offering the widows only $5,000 each. "The indifference of these companies to suffering," he exclaimed. "I can't take that."

Stern, therefore, brought a "wrongful death" suit for $60 millions against the Blue Diamond Coal Corporation. The first trial ended in September, 1977, with a victory for Blue Diamond. The judge, H. David Hermansdorfer, ruled—without allowing a jury to study the case's merits—that workmen compensation protected the coal company "from suits by employees or their survivors." (Workmen compensation came to roughly $400 a month for each widow and her family).

The suit, Jennifer Boggs, et al, vs. Blue Diamond Coal Company, was appealed, and was scheduled for trial on August 12, 1980, in Frankfort Kentucky. Shortly before it was to begin, however, the case was settled out of court for approximately $5.9 millions. Judge William O. Bertelsman said that a "properly instructed jury could find that the evidence in this case would warrant a finding of liability against the Blue Diamond Coal Company." He added, however, that Blue Diamond's payment did not constitute "an express or implied admission of liability in this or any other

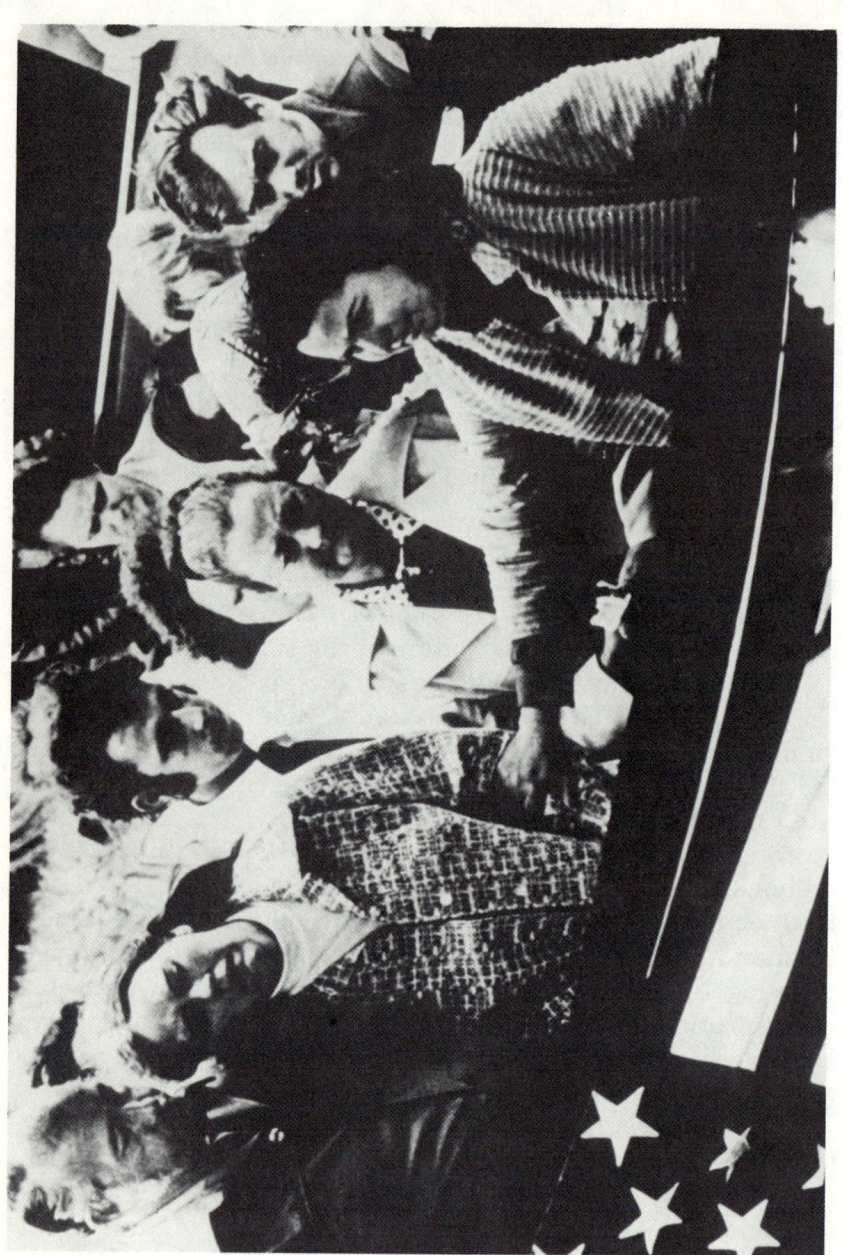

Scotia, November, 1976. Photo courtesy, UMWA.

decision."

Stern responded to this statement about liability by saying, "They [Blue Diamond] don't pay [nearly] $6 millions...if they aren't liable." Reda Turner Lawson, whose husband was lost in the explosion, said "victory over the company was what we wanted, and I think it's obvious that if they agree to pay us, it proves their guilt." Lawyer Combs, however, said the suit against his client was sapping corporate resources "that should have been devoted to mining coal," a situation that greatly encouraged the company to settle.

Mrs. Boggs received $50,000 immediately. Then she would get $602.69 a month for life, with a minimum guarantee of fifteen years; her young son, Ivan, would receive $802 a month until his majority. This settlement was regarded as typical for the other fourteen widows. It was far different from the money the widows and children would have received under Kentucky's workmen compensation laws.

Stern said of the settlement that he hoped "recovery from a parent corporation in spite of workmen's compensation will cause coal companies in general to be more concerned about safety. That's the point."

In the meantime, Scotia and Blue Diamond girded themselves for defense against a six-point federal indictment of June, 1979, charging them with offenses that ranged from bad ventilation to dangerous accumulations of methane and coal dust. And so it goes.

The Scotia-Blue Diamond case does not herald a "new day" for those left behind by coal-mine accidents. Settlements will be made—if they are made at all—case by case, with few or no precedents allowed. Survivors will still have to go through the legal mazes of "workmen compensation," and unfriendly judges, and high-powered attorneys. The Scotia settlement, however, did indicate that there are lawyers now willing to spend years, if necessary, to present survivors' sides of the story. In this respect, therefore, Scotia could very well be an index to the future.

Little good, though, that does a Letcher County, Kentucky family named Callihan. They lived near an abandoned strip mine, where work ceased in the early seventies. In July, 1980, tons of mud, rocks, and trees tumbled from the old workings down the mountain, stopping just short of the Callihan house. The avalanche left a huge pile of debris in front of their residence of twenty years, cutting off all motorized access. Another fall of at least 19,000 cubic yards is expected.

The company that owned the mine is in Virginia, and no longer does business in Kentucky. Thus, say Kentucky officials, the company is not liable. The State, however, disclaims any responsibility either for clearing up the first fall, or preventing another one. The County Magistrate did have a temporary, good-weather track built through the fall, but said, "There's way too much dirt for us to move. And besides, the County don't own the road...." Everybody wants the federal government's Office of Surface Mining to take care of the problem. And that means "feasibility studies," and "engineering evaluations." And those take time. And, meanwhile, another slide comes closer, and closer. And so it goes.

Sources

Interview: Gerald Stern, September 18, 1980.

Interview: Frank O'Gorman (U.S. Mine Safety and Health Administration), September 16, 1980.

Interview: Leonard Gross (Consolidation Coal Corporation), September 17, 1980.

Louisville *Courier-Journal*. November 22-3, 1968; August 17, 24, 1980.

New York *Times*. November 21, 1968.

Park City Daily News (Bowling Green, Kentucky). November 21-2, 1968; November 24-30, 1968.

United Mine Workers Journal, No. 11, June 15-July 15, 1976.

Epilogue

Fie on an industry that killed nearly a thousand of its employees a year, while all the time resisting change that might possibly have made it otherwise. Fie, too, on a government that allowed the industry to get away with it. One does not have to be a "socialist" or a "leveler" to condemn such practices.

Beyond the fatalities, one must note that the 20th century in U.S. coal mining is literally strewn with bent-over, crippled figures; with others wheezing and choking their way through attacks of "black lung." Truly it is so, as one miner put it: "God made the coal and He hid it. Then some fool found it, and we've been in trouble ever since." These then, are the first and abiding images that come from the dreadful month of December, 1907: the filled-up graveyards, and the enduring pain and suffering of the survivors.

Another image that these events conjure up is a public caught between hero worship and popular democracy. Throughout most of our history, we have revered certain individuals who have risen above the system—and for a long time the industrialist held that place. In the 1890s, if asked what he wanted to be when he grew up, a youth would frequently answer "a businessman," or "an industrialist," an attitude brought on in no small measure by writers like Horatio Alger, whose forte was "rags to riches" stories. By 1910, however, the glamor of the "Captain of Industry" was, for various reasons, somewhat faded. Thus, the great public, remembering the heroes, but now knowledgeable of the need for industrial humanitarianism, was torn between the *status quo* and a general reform movement. Generally, this balancing act by the public gave the big corporations the lee-way they wanted to have their way about things.

When the slaughter reached such spectacular levels that the public could not bear the publicity, it began to desert those paragons of "individualism," the corporate founders, and come down hard for industrial democracy. Or, the same result could sometimes be reached if the public ever really became aware that it was lied to so much. This, then produces a third image of the dreadful month: the ultimate propensity of the public—in a crisis—automatically to disbelieve anything said by a company or government official.

This still goes on today. Note the nuclear emergency in April, 1979, at Three Mile Island, Pennsylvania. Practically the first words of a utilities spokesman were: "It doesn't look like a very serious accident at this point." Another official followed this with "There is absolutely no danger of a melt-down. We are not in a China Syndrome type situation." A state government official said at first, "Everything is under control," only to lose his temper five hours later because he felt the utility company had misled him.

In situations like this—where spokesmen tell themselves one thing, and the populace another, what is the general public to do? For a great many years it simply shook its head, and smiled ruefully at the "big-shots" who were unable to reach any firm decisions. But that was before the days of radioactive fallout. Now it is, in some people's minds, a matter of human survival in general. Thus, the "credibility gap" that has been building for so many years appears now to be at its height. Surely, the false reports and the erroneous impressions that emanated from so many coal mine disasters helped this mistrust to develop.

For its part, the government has evolved its own "credibility gap." Each time it is confronted by groups of protestors, it greets them as "ignorant trouble-makers," refusing to pay serious attention to their ideas, because that would too frequently be a refutation of its own "experts." But it has happened so many times that the "unreasonable" and "illogical" protests have ultimately become the policy. One thinks immediately in this respect of Vietnam and environmental protection.

Maybe there should be created a "Bureau of Minds," drawn from the public and private sectors, which could somehow bridge the gap between government-business and the public interest. The distrust has become increasingly obvious from the coal-mine tragedies of 1907 and the following years to the nuclear fears of the present. It is easy, as seen by past experiences, to become extreme in either direction; perhaps, then, an "ombudsman," or "liaison" agency would produce the moderation that seems so elusive during times of crisis.

It is a well known fact that an extreme action always invites an extreme reaction. Many people—and in some instances, rightly so—lament the unusual amounts of power that labor unions, the UMW included, have today. The biggest reason, however, for the unions' power now was the intransigence of management then. This, then, provokes a fourth image that grew out of the dreadful month: powerful unions—sometimes so powerful that they could actually dictate to the government—an unseemly quality, to say the least.

But one cannot make a study of laboring conditions in this country at the beginning of the twentieth century without developing some kind of empathy for the common laborer. One sees why change was inevitable. The system of "industrial feudalism" simply could not last. Some other direction, it was clear, would have to be taken. And if that journey wound up at a destination called "Union Power," it is because the trip started at "Management Arrogance."

There are certain questions in this country which have usually brought the charge of naivete to their inquirer:

. Why couldn't management offer its labor decent wages, decent housing, and decent working conditions? Why should it become involved in practices that are so obviously inimical to human compassion?

. Why couldn't unions yield at least to some degree in the interest of industrial harmony?

These questions are not naive as much as they are basic. Unfortunately, they have not been asked enough, and this brings up a fifth image from the dreadful month: the dialogue between labor and management turned almost completely to that of confrontation; rarely compromise, unless the government, acting under instructions from the public, forced them to.

Why should all these deaths, injuries, illnesses, fears, and heart-breaks come to nought? Why should it simply be "business as usual" in the struggles between labor and management, and the mutual distrust between government-business and public opinion? Would it not be wonderful if all those industrial sorrows were melded together to act as a symbol by which industrial progress was accomplished? That way, those people certainly would not have died in vain. That way, too, perhaps a bit of harmony could replace the almost automatic discordancy that crises produce in the various factions of our society. After all, each of these—labor, management, government, public opinion—do have elements that are deserving of trust. And that is a point worth pursuing.

Index

CIVC

SUNDAY SERVICE

SEPTEMBER THROUGH MAY
FROM 1:00 P.M.-5:00 P.M.
IN CHARLESTON, DUNBAR AND ST. ALBANS

Library materials may be returned to
any of the public libraries in Kanawha County.

After hour book drops are
available at all locations.